Celebrated biologis... a worldwide field trip... covers a single frog d... how twenty-four different species spend their time. In the early hours of the morning, we hear a horned marsupial frog "bopping" and a wood frog "quacking" to attract mates. At six o'clock in the morning, beneath a streetlight in Honolulu, we meet a corpulent, invasive cane toad slurping insects—and sometimes snakes, lizards, turtles, birds, and mice. At noon, we watch parenting in action as an African bullfrog bulldozes a path through the mud to free his tadpoles from a drying pond. At dusk, in a Peruvian rainforest, we observe "the ultimate odd couple"—a hairy tarantula and what looks like a tiny amphibian pet taking shelter in the spider's burrow.

For each hour in our frog day, award-winning artist Tony Angell has depicted these scenes with his signature pen and ink illustrations. Working closely together to narrate and illustrate these unique moments in time, Crump and Angell have created an engaging read that is a perfect way to spend an hour or two—and a true gift for readers, amateur scientists, and all frog fans.

Praise for **FROG DAY**

"Crump is a world expert on the private lives of frogs and toads. She has spent decades as a professional Frog Voyeur, discovering astonishing complexity in how frogs overcome daily challenges. This delightful book will surprise the reader with tricks that frogs use to survive and breed and open the reader's eyes to the marvelous world of small hopping creatures."

RICK SHINE, author of *Cane Toad Wars*

"*Frog Day* is packed with fact-filled vignettes spanning these spectacular amphibians' global occurrence, sizes, appearances, habitats, behavior, and conservation status. Readers of this wonderful book will learn a lot about frogs, as well as why we should all be concerned for the plight of wild nature on our rapidly changing planet."

HARRY W. GREENE, author of *Tracks and Shadows: Field Biology as Art*

"We truly could not have asked for a more knowl-
edgeable and enchanting guide. As we hop from
continent to continent, we not only get a glimpse
of the natural history, behavior, and adaptations
of frogs, but we also see how these creatures are
intertwined with culture and folklore from around
the world. As *Frog Day* comes to an end, we are
left with a sense of wonder and a thirst to
continue this exploration."

SINLAN POO, coeditor of *Women in Herpetology:
50 Stories from around the World*

"In rainforests around the world, pesky mosquitoes swarm your face while frogs 'fly' through the
canopy, call from branches dripping with ferns and
orchids, and mimic plump, ripe tomatoes. Elsewhere, crawfish frogs snore, fat cane toads gobble
up insects, and African bullfrogs babysit their
tadpoles. Readers will delight in this beautifully
illustrated and evocative reminder of how special
amphibians and their habitats are."

SUSAN C. WALLS, United States Geological Survey

"This is storytelling at its best. *Frog Day* is an engaging and visually stunning narrative on frogs—some of the planet's most imperiled creatures. Crump has skillfully woven together tantalizing tales with fascinating facts, resulting in a beautiful celebration of frogs at a time when they need our help the most. I have no doubt that this book will captivate readers of all ages, at the same time inspiring a passion for frogs and their conservation."

JODI ROWLEY, the Australian Museum & UNSW Sydney

Frog Day

FROG
DAY

A STORY OF 24 HOURS AND
24 AMPHIBIAN LIVES

Written by	Illustrated by
Marty Crump	Tony Angell

The University of Chicago Press
Chicago and London

The University of Chicago Press, Chicago 60637
The University of Chicago Press, Ltd., London
© 2024 by The University of Chicago
Illustrations © 2024 by Tony Angell
Published 2024
Printed in China

33 32 31 30 29 28 27 26 25 24   1 2 3 4 5

ISBN-13: 978-0-226-83020-9 (cloth)
ISBN-13: 978-0-226-83021-6 (e-book)
DOI: https://doi.org/10.7208/chicago/9780226830216.001.0001

Library of Congress Cataloging-in-Publication Data

Names: Crump, Martha L., author. | Angell, Tony, illustrator.
Title: Frog day : a story of 24 hours and 24 amphibian lives / written by Marty
    Crump ; illustrated by Tony Angell.
Other titles: Earth day (University of Chicago. Press)
Description: Chicago ; London : The University of Chicago Press, 2024. |
    Series: Earth day | Includes bibliographical references and index.
Identifiers: LCCN 2023057803 | ISBN 9780226830209 (cloth) |
    ISBN 9780226830216 (ebook)
Subjects: LCSH: Frogs. | Frogs—Pictorial works. | LCGFT: Illustrated works.
Classification: LCC QL668.E2 C876 2024 | DDC 597.8/90222—dc23/eng/20240125
LC record available at https://lccn.loc.gov/2023057803

♾ This paper meets the requirements of ANSI/NISO Z39.48-1992
(Permanence of Paper).

Contents

6 AM
Cane Toad
(SOUTH AMERICA,
INTRODUCED ELSE-
WHERE)
39

7 AM
Common
Midwife Toad
(EUROPE)
45

8 AM
Golden Poison Frog
(SOUTH AMERICA)
53

9 AM
Pacific Horned Frog
(SOUTH AMERICA)
59

10 AM
Tomato Frog
(MADAGASCAR)
67

11 AM
Bornean
Foot-Flagging Frog
(ASIA)
73

NOON
African Bullfrog
(AFRICA)
79

1 PM
Rockhole Frog
(AUSTRALIA)
85

2 PM

Giant Frog

(AUSTRALIA)

91

3 PM

Yucatán
Shovel-Headed
Treefrog

(MEXICO,
CENTRAL AMERICA)

97

4 PM

Oriental
Fire-Bellied Toad

(ASIA)

103

5 PM

Darwin's Frog

(SOUTH AMERICA)

111

6 PM

Reticulated
Humming Frog

(SOUTH AMERICA)

117

7 PM

Rosenberg's
Gladiator Frog

(CENTRAL AND SOUTH
AMERICA)

123

Preface

What do frogs *do* all day long? Join me on a twenty-four-hour field trip to discover for yourself. We'll peer into the lives of frogs as they survive harsh environments; eat ants, mice, and each other; defend territories and attract mates; protect themselves with poisons and sticky glue stronger than rubber cement; and care for their young in pouches, in vocal sacs, and wound around Dad's hind legs. We'll travel to six continents— everywhere frogs are found. We'll tolerate sweating and shivering and brave mos- quitoes and leeches for the opportunity to watch frogs perform amazing tricks. Each chapter highlights one species of frog at a particularly exciting time of its day.

As your guide on this trip, allow me to

introduce myself. I'm a herpetologist—a biologist who studies amphibians and reptiles. For over fifty-five years, I have studied the ecology and behavior of frogs in Costa Rica, Brazil, Ecuador, Argentina, and Chile. My specific interests include amphibian conservation and the diversity of ways frogs reproduce and care for their young. We know little about the natural history of thousands of species of frogs, especially those that occur in remote areas of the world. One of the thrills of being a field biologist is discovering previously unknown behaviors. That possibility is always waiting just around the next bend in the trail.

But to make those discoveries, we need to know where to find frogs. As I write this sentence in August 2023, there are 7,642 recognized species of frogs. (The number of frogs worldwide changes as new species are described and others are lumped together or go extinct.) Frogs live on every continent except Antarctica, but they are not distributed evenly around the globe. The tropics are home to the greatest number of species by far, and the tropical region that extends from Mexico through Brazil, encompassing the vast Amazon rainforest, has many more species than the tropical regions of Africa, Asia, and Australia. Tropical South America is the world's richest area for frogs. For example, Brazil has 1,178 native species; Peru, 649; and Ecuador, 642. In contrast, consider some temperate areas: the United Kingdom has only 4 species of native frogs; Egypt, 11; Germany, 13; Canada, 24; Russia, 24; the United States, 106; South Africa, 132; and Argentina, 171. This uneven distribution explains why our field trip is geographically skewed; we'll spend more time on South

America than on other continents because that's where the frogs are.

Frogs live in a wide variety of habitats, from deserts to rainforests and from coastal areas to mountaintops. Not surprisingly, then, frogs are remarkably diverse in appearance and behavior. They range from the size of a pea to that of a large dinner plate. They come in all colors, including pink, purple, blue, and lemon yellow. And they can be solid, striped, or polka-dot. Most species abandon their eggs in water, and the hatched tadpoles—swimming blobs with tails—eventually metamorphose into four-legged, tailless adults. Other frogs lay their eggs on land and are devoted parents; many of these species skip the tadpole stage altogether. Some frogs burrow underground, some live high in trees, and some spend their entire lives in water. Although frogs quickly lose water through their skin, species from diverse families have mastered life in the desert. Frogs serve as both predators and prey; they're an integral part of our ecosystems. The twenty-four species highlighted here illustrate a minuscule fraction of the group's rich diversity.

We will meet some *frogs* and some *toads*. What's the difference? In scientific terms, toads belong to the family Bufonidae. This means that all toads are frogs, but not all frogs are toads. (Just as all dachshunds are dogs, but not all dogs are dachshunds.) In common usage, however, the term *toad* generally refers to a frog that has drier skin, a squat and often warty body, and hind legs that are not particularly powerful. Toads tend to flop when they jump, in contrast to the Olympic-level leaps of other frogs. Two of the species we'll encounter on our field trip, the common midwife toad and the Oriental fire-bellied toad, are not members of the family Bufonidae and are therefore not toads in the scientific sense. Their common names reflect their warty bodies. The true toads we'll meet are the cane toad and the common toad.

Populations of frogs are declining—and even disappearing—all over the world. Of the 7,642 recognized species of frogs, more than 500 are considered critically endangered by conservation biologists. Some of those species may now be extinct, adding to the thirty-two confirmed extinctions within the past several decades. Examples of species we've lost include the golden toad from Costa Rica, the splendid poison frog from Panama, and the southern gastric-brooding frog from Australia. Scientists identify six main causes for frogs' declines and disappearances:

habitat destruction or modification, disease, overexploitation, pollution, introduced species, and climate change. We'll meet some of the threatened species on our field trip.

Humans have, from our very beginnings, used art and story to understand and describe our natural surroundings. By combining science with narrative and Tony Angell's illustrations, I hope to make frog natural history come alive—to hatch a tadpole of interest that will metamorphose into a frog of deep appreciation.

MARTY CRUMP
Logan, Utah

Artist's Note

Portraying, in pen and ink, a wide range of our wild cohorts in nature has taken me on adventures of discovery. My work with frogs is no exception. This particular adventure began in the 1940s, when as a kid I would scoop up pollywogs from a pool adjacent to the Los Angeles River, portions of which were then still running free. In the summer, my family would visit Michigan, where my uncles and I caught bull frogs in the lake by their cabin. I remember once scurrying breathlessly through the woods to follow a bull snake as it chased a leopard frog that had wandered too far from the safety of its pond.

While I look back fondly on those moments, the opportunity to illustrate *Frog Day* has imparted an understanding of and appreciation for the species' diversity that extend far beyond my initial experiences. Their remarkable strategies for survival, from forming mutually beneficial partnerships with very different animals to employing unusual brooding behaviors, leave me in awe. Add the elegance of their forms, patterns, and colors, and you have a compelling subject for study.

All life on Earth faces climate change and the intensifying demands on—and damage to—our environment, but frogs are particularly vulnerable to these threats. At the same time, they continue to tell us how their subtle role within an ecosystem is essential to the vibrancy of its complex tapestry. Are we listening?

TONY ANGELL
Seattle, Washington

Frog Day

MIDNIGHT
Wallace's Flying Frog

Rhacophorus nigropalmatus

(ASIA)

Look up. It's a bird! It's a plane! No, it's Wallace's flying frog—a lean, green flying machine—whizzing through the palm trees.

The exuberance of life surrounding us in this Indonesian rainforest heightens our awareness. We smell the soil, night-blooming flowers, and decomposing vegetation. We hear the buzzes, chirps, and peeps of insects and frogs. We feel the moisture. The Wallace's flying frog that just whizzed by, caught in our headlamp beams, startled us. But it's more frightened than we are; it's fleeing from a hungry tree snake.

But flying? Well, not really. Frogs can't fly any more than pigs can. Wallace's flying frogs, however, have perfected the art of gliding. These nocturnal frogs spend most of their time high up in the trees, descending to the forest floor only to mate and lay eggs in wallows made by large mammals or puddles formed by heavy rains. When disturbed, they leap from their perches, navigating the canopy by splaying their limbs sideways and banking. Thanks to the skin flaps along their legs, their oversized feet, and the full webbing between their toes, they can glide over fifty feet (about fifteen meters) to other branches. The extensive webbing slows their descent and helps them steer, while the large pads on the tips of their toes ensure a soft landing.

The common name of this species honors Alfred Russel Wallace, the nineteenth-century English naturalist and biologist who independently proposed the theory of evolution through natural selection at the same time as Charles Darwin. Wallace formulated his ideas while exploring the Malay Archipelago in the 1850s. It was there, in Sarawak on the island of Borneo, that he collected the first recorded specimen of this frog.

Flying frogs appear in various folktales from around the world. Be glad we were startled by Wallace's frog in Indonesia rather than the *llamhigyn y dwr* in Wales. Welsh folklore tells of this evil giant frog-like creature with bat's wings, a serpent's tail tipped with a wicked stinger, and two needle-sharp fangs. The *llamhigyn y dwr* inhabits swamps, rivers, and lakes, soaring through the air in search of prey that venture too close to its water: sheep, cattle, and unsuspecting fishermen.

Whereas the mythical *llamhigyn y dwr* is a dull greenish brown, our avocado-sized Wallace's flying frog is a vibrant green with unusually large, pale-yellow eyes. But it wasn't always so bright; plant pigments called carotenoids get the credit for that. Just as humans get immune-boosting antioxidants from eating plants high in carotenoids (such as carrots, yams, and tomatoes), other animals get their colors. Flamingos, for example, are pink or peachy orange because the algae and tiny crustaceans they eat are rich in carotenoids. Wallace's flying frogs are orange red when they first metamorphose from tadpoles but turn lime green as they mature because of the carotenoid-loaded insects they eat.

Is there an advantage to being orange red rather than green when young? The tiny froglets resemble the orange-red droppings of the Asian glossy starling (*Aplonis panayensis*) and several species of fruit bats. This might be a lifesaving adaptation, as it makes the juveniles unrecognizable and unappetizing to birds and other predators. But when the frogs grow four times in size, the poop color no longer fools anyone.

Despite a popular song's lament that it's not easy being green, our adult Wallace's flying frog finds safety in this color, which lets it blend in with the surrounding foliage. But even if a predator detects and approaches the frog, it can simply splay its limbs, spread its toes, and glide to a neighboring tree.

Horned Marsupial Frog

Gastrotheca cornuta

(CENTRAL AND SOUTH AMERICA)

A loud "bop," reminiscent of the popping of a champagne cork, breaks the silence. Ten minutes later, another "bop."

We shine our headlamp beams upward and trace the sounds to a branch festooned with moss, ferns, liverworts, and lavender and white orchids deep in this Ecuadorian rainforest. About midway down the branch sits a caramel-colored horned marsupial frog. These frogs spend most of their lives higher in the forest canopy, so we're fortunate to get a good look at this handsome fellow, who wears hornlike flaps of skin above his bronzy-gold eyes.

The champagne might be uncorked, but it isn't time to celebrate yet. Our bopping frog is calling to attract a female. If he is successful, at least two hours of mating gymnastics will ensue. His mate will release up to a dozen eggs, one at a time. After fertilizing the first egg, he will use his hind feet to push it through the opening in the brood pouch covering her back. She will release another egg, and the routine will continue until all fertilized eggs are secured in her pouch.

A female horned marsupial frog does not produce many eggs, but each one is energy rich—and unusually large. In fact, at about the size of a thumbtack head, this species' eggs are the largest of any known frog. A female will brood her eggs for two months, until tiny froglets hop out from their incubator, skipping the tadpole phase found in most frog species. This process is called "direct development." The embryos have external bell-shaped gills that spread out against the inside wall of their mother's pouch, allowing for efficient gas exchange. Consider them "gill placentas!"

Scientists have long assumed that young direct-developing marsupial frogs get all the nutrients they need from their massively yolked eggs. Researchers have recently discovered, however, that female Abra Acanacu marsupial frogs (*Gastrotheca excubitor*) "feed" their developing young: nutrients from their insect meals are incorporated into their offspring, presumably through the pouch membranes.

Time may be limited to investigate the

many mysteries of horned marsupial frogs. They are disappearing, and they are not the only ones. We are threatening an estimated 41 percent of amphibians with extinction. While disease and climate change take a general toll, it is habitat loss and modification specifically that threaten more than one-third of these species. Humans degrade forest habitats by logging, building houses, and farming. These activities remove plants and animals, introduce pathogens and predators, and change food resources.

Horned marsupial frogs have disappeared from Costa Rica and western Panama, and they are now declining in Colombia. When the species seemed to have disappeared from Ecuador in 2005, scientists feared that the country had forever lost these unique frogs. But in 2018 biologists discovered six individuals and heard more bopping from the canopy in a remote part of the Choco rainforest—the same forest we're exploring during this early morning hour. The Ecuadorian Choco experiences the highest rate of deforestation in the country, but thankfully, this section of forest is protected. At least here, horned marsupial frogs have a good chance of persisting. Our bopping male is leading the charge.

2 A M

Mud-Nesting Frog

Leptodactylus bufonius

(SOUTH AMERICA)

Last year's roadside ponds have receded to mere puddles. The air is dry, and the night sky is cloudless. A male mud-nesting frog sits beside what resembles a mini volcano at puddle's edge, belting out his metallic "huinc, huinc" in the hope of attracting a mate.

On this ranch in Joaquín V. González, in the Salta Province of northern Argentina, a pig wallows in a shrinking pond. Goats munch on thorny vegetation. Swirling dust and dirt blow into our eyes and nostrils. It's the end of the dry season, and everyone seems to be waiting for rain. Mud-nesting frogs have conquered this scrub forest and are, in fact, one of the most common frog species in the hot and semiarid lowland area known as the Gran Chaco in Argentina, Bolivia, Brazil, and Paraguay.

Mud-nesting frogs need standing water to reproduce. So, what makes them so well adapted to this harsh environment, where both the timing and amount of rainfall are unpredictable? Their mud nests. The male begins nest construction by selecting a spot near the edge of a drying pond and rotating in a circle until he has pushed the mud away from his body. Then, using his shovel-shaped snout, he forms the mud into a cone about 4.3 inches (11 centimeters) high with an opening at the top of about 1 inch (2.5 centimeters)—just wide enough for the frog to crawl through. After he has finished his nest, he takes the evening off. When he returns, he sits either inside or near the volcano and calls to attract a female—"huinc, huinc."

A portly female approaches our hopeful male. The two frogs engage in a brief courtship, calling softly back and forth to each other. Our male then climbs into his mud nest. The female follows. While she lays her clutch of five hundred eggs, he uses his hind legs to whip eggs, mucus, and sperm into a frothy white foam. Once finished, he climbs out of the nest. Again, she follows. The male hops away and disappears under a spiny bush, but the female has one last parental chore. With her front feet, she pushes mud along the sides and onto the top of the nest to plug the opening. Sealed with care, the mud nest will keep the fertilized eggs moist, cool, and safe from predators.

Once hatched, the tadpoles stay inside their mud nest until rains wash them into a newly formed pond. Enter unpredictability. The rains may come in a few days—or many weeks later. Most aquatic tadpoles excrete ammonia, a fairly toxic waste product. This isn't a problem for tadpoles that swim in pools or ponds. Their ammonia gets diluted, and the tadpoles don't poison themselves. Mud-nesting frog tadpoles don't have such spacious accommodations, and yet they can survive inside their mud nest for over a month. How? By instead producing mostly urea. Urea, also found in human urine, is less toxic than ammonia—a great adaptation to keep the tadpoles' muddy homes from becoming muddy tombs. The tadpoles thus maintain a relatively pleasant place to live until they are freed by the rain and find themselves swimming in a pond with food to eat. They are not home free, however. They must now dodge predators, including carnivorous "pacman" tadpoles. (We'll meet these in a few hours.)

Mud-nesting frogs have adapted to their harsh environment in yet another way: they borrow the homes of other animals. Vizcachas, rabbitlike rodents belonging to the same family as chinchillas, construct their burrows within tangles of thorny vegetation. Mud-nesting frogs are sometimes called "vizcacheras frogs" because they shelter inside those burrows, seeking refuge from the baking sun and protection from hungry snakes. They may also spend their daylight hours in abandoned tarantula burrows.

This unique combination of

adaptations—mud nests, urea, and bor-
rowed homes—enables mud-nesting frogs
to thrive on this ranch in the seasonal Chaco,
along with the pigs, goats, vizcachas, and
tarantulas.

Wood Frog
Rana sylvatica
(Lithobates sylvaticus)

(NORTH AMERICA)

It seems too early in the year for frogs to be calling their hearts out, but there by a patch of melting snow sits a male wood frog— "quack, quack."

Early spring has brought warmer temperatures and snow melt, but we still shiver in this moist Canadian woodland dominated by oak and maple trees, listening to the soft duck-like cackling of wood frogs emerging from hibernation. The frogs have

spent the winter frozen under the leaf litter—no heartbeat, no breath, no muscle movement—and have now thawed back to life.

The wood frog's liver produces glucose. As winter sets in, this sugar flushes into the body's cells, preventing them from freezing. Consequently, as ice forms between the frog's layers of skin and muscle and fills the animal's abdomen, it encases the internal organs but crucially does not enter the cells. Researchers are studying this glucose antifreeze. If hospitals can master the wood frog's trick, doctors might be able to freeze and thaw human organs used for transplants without damaging them. And that could save lives.

Wood frogs are brown, reddish brown, gray, or tan and have a distinctive dark-brown or black mask across their eyes. They

are the only frogs in the Western Hemi-
sphere to live above the Arctic Circle. They
breed in ephemeral pools and ponds and are
one of the first frog species to lay eggs once
the snow melts.

Snow cover is critical for wood frogs: it
insulates the ground where they hibernate
and thereby keeps the temperature relatively constant. Fluctuations in ground temperature would cause wood frogs to experience multiple freeze-thaw cycles during the
winter months. And each time a frog thaws,
its warmer body temperature requires more
energy. If the frogs use too much energy
just to maintain their bodies, the males may
not have enough to call to females and the
females may not have enough to mature
their eggs. Climate change, with its effects
on patterns of precipitation, poses a real
danger to wood frogs. If snowpack is less

than usual and the frogs thaw even once too often, they might begin the breeding season with insufficient energy reserves.

But that's not the only problem wood frogs might encounter in our changing world. Global warming is affecting the seasonal timing of plant and animal activities everywhere. Mammals are emerging from hibernation, plants are blooming, and birds are nesting earlier and earlier. Warming conditions, especially at high latitudes, are causing male wood frogs to call and female wood frogs to lay their eggs too soon in the season. How many of those fertilized eggs will hatch into tadpoles and how many of those tadpoles will metamorphose into froglets? The pond could freeze over again, killing all the eggs. If the eggs do hatch, the tadpoles will likely experience colder water temperatures, leading to slower development. And more time spent in the pond means more time spent with predators, such as insect larvae, salamanders, and turtles.

As if those problems weren't enough, runoff of salt used for deicing roads harms wood frogs. Tadpoles exposed to road salt become emaciated, develop bent tails, swim in circles, don't respond quickly to predators, and often die. If they metamorphose, frogs exposed to road salts often experience reduced growth rates, reproductive success, and survivorship.

Wood frogs can conquer harsh winter conditions thanks to their glucose antifreeze, but can they survive a warmer, saltier world?

Wandolleck's Land Frog

Sphenophryne schlaginhaufeni

(NEW GUINEA)

A brown blob hops in the damp leaf litter. A closer look reveals a frog with a passel of froglets clinging to both sides of his body. Where are they headed?

About three-quarters of the length of your pinky finger, this Wandolleck's land frog fertilized his mate's twenty eggs under the damp leaf litter. His mate took off, but

he stayed with the eggs day and night, hunkering over them to keep them moist and lunging at beetles, ants, and spiders that ventured too close. Within three days, all twenty of the eggs hatched into tiny froglets, bypassing the tadpole stage, then climbed onto the sides of their father's small body.

During the day, this devoted dad and his offspring hide under leaf litter. Then, under cover of darkness, he takes his passengers on a ride through the forest, traveling up to fifty-five feet (about seventeen meters) each night for up to nine consecutive nights. The froglets jump off individually in different areas of the forest, on different nights, and at different times. They ride until something stirs them to take that leap to independence, at which point they crawl onto the middle of their father's back, advance forward, dive over his head, and tumble into the leaf litter.

What's the advantage of this unusual parental transport? The behavior distributes the froglets in the forest, potentially reducing their competition for food, decreasing their risk of predation, and lowering their chance of inbreeding. It works because the forest is wet, allowing the male and his direct-developing offspring to stay moist.

Parental behaviors that increase the chances offspring will survive—what scientists call "parental care"—occur in about 20 percent of the more than 7,600 species of frogs. Most of these species deposit their eggs on land, where parental care helps keep the eggs moist. Only one other frog is known to transport froglets hatched from terrestrial, direct-developing eggs: the Jamaican cave-breeding frog (*Eleutherodactylus cundalli*). In this species' case, the female sits on her eggs in the natal cave for a month or so. A cave provides a relatively safe environment for eggs, but its food supply for growing juveniles—particularly for up to seventy-two of them—is scarce. So, once the eggs hatch, the froglets climb onto their mother's back. She then carries them out of the cave to where food is more abundant.

Has witnessing this amazing behavior been worth offering up your blood to leeches in this humid rainforest? To whining mosquitoes attacking your face, neck, and hands? Has it been worth getting soaked in the constant drizzle and sinking in mud up to your shins? Absolutely. Not many people have glimpsed this spectacle of froglet transport.

Crawfish Frog

Lithobates areolatus
(Rana areolata)

(NORTH AMERICA)

The plump, egg-laden crawfish frog clumsily hopping over a grass clump is on a mission. Her long trek from home, through this prairie of big bluestem, Indian grass, and blazing star, to the breeding pond is almost over.

Crawfish frogs get their common name because they live in the abandoned burrows of several species of upland crayfish—or crawfish, or even crawdads, depending on where you come from. These crustaceans live on land and burrow about three feet (one meter) deep, down to where the soil is saturated with water. In time, they abandon their burrows for fresh ones, leaving the old to crawfish frogs, whose bodies are conveniently about the same width as the burrows. These secretive frogs live in their moist homes for more than ten months of the year. In northern regions, the frogs overwinter at the bottom of these burrows, in or near the water level. In warmer southern areas, the frogs spend less of the winter inactive but still, like their northern relatives, rarely venture far from their homes.

Crawfish frogs generally stay next to their burrows on "feeding platforms," from where they ambush unsuspecting prey—ants, beetles, and crickets; centipedes and millipedes; even their homes' architects, crayfish. When the frogs are on their feeding platforms, they face the burrow entrance, ready to dash to safety if necessary. Once in the burrow, they turn around, face the entrance, fill their lungs with air to inflate their bodies, and lower their heads. Snakes and other predators are hard-pressed to dislodge such ensconced frogs.

But because there is no water nearby for breeding, the frogs must leave their burrows in the springtime and migrate to wetlands in prairies and pastures. These migrations occur at night and into the early morning hours. Male crawfish frogs recently arrived at the shallow ponds in the prairie we are now visiting and have begun uttering their loud, deep snores—"gwwaaa, gwwaaa, gwwaaa." They sound rather like ravenous pigs at feeding time. Both males and females follow familiar routes to breeding ponds and back home again—an average of nearly 0.3

miles (0.5 kilometers) each way, although some frogs travel over twice that far.

It takes crawfish frogs, on average, about five days to travel to their breeding sites and at least thirteen to return home to their burrows. The outward-bound trip likely takes less time because of the frogs' motivation to breed and because the raucous snores of the early arrivals guide females and other males to the sites. Migration is hazardous. The frogs are traveling when hungry snakes have emerged from hibernation and are searching for food. Eastern hog-nosed snakes, eastern garter snakes, and black racers—not to mention raccoons—all view crawfish frogs as delectable prey. Migrating crawfish frogs have another problem. Human development of the frogs' prairie and grassland habitats means that the frogs sometimes need to cross roads during their travels. And they don't always get to the other side.

If migration is so hazardous, why do crawfish frogs journey so far? Of course they need to get to a wetland to breed, but why do they return to the same burrows? Is there no acceptable habitat in between? Perhaps crawfish frogs simply don't want to let a good thing go.

6 AM

Cane Toad
Rhinella marina

(SOUTH AMERICA, INTRODUCED
ELSEWHERE)

A corpulent cane toad sits beneath a residential streetlight, slurping insects—flying termites, beetles, crane flies, and moths—as they fall to the ground. We're standing on a corner in Honolulu, Hawaii, watching this 3-pound (1.4-kilogram) glutton fill her belly. Cane toads are not native to Hawaii, but they have made themselves at home in the state.

Cane toads have become international travelers thanks to their voracious appetites. Humans introduced these toads into many areas of the world in the hope of controlling insect pests in sugarcane fields. People brought cane toads from French Guiana to Puerto Rico, then introduced their descendants to Hawaii in the 1930s. But cane toads eat more than agricultural pests. They eat just about anything they can stuff into their cavernous mouths. Their diet includes snails, insects, small snakes, lizards, turtles, nestlings, and mice. A toad will spend hours picking off flies swarming around table scraps on a compost pile or pet feces on a lawn. And they don't stop with live prey. They sit down for dinner in people's backyards, eating any dog or cat food left outside.

Homeowners despair when these brown, warty toads the size of soup bowls invade their neighborhoods. Why? Because cane toads secrete a white, milky poison from large triangular glands on the sides of their heads. Eating a cane toad just might kill a dog or a cat.

Scientists call nonnative species that are likely to cause economic or environmental harm "invasive." Cane toads are considered one of the worst invasive vertebrates in the world. They can live in a wide variety of open, disturbed habitats, like riverbanks, treefall gaps in forests, and human-populated areas. They can move far and quickly—and as we have seen—are not picky eaters. A single clutch might contain thirty thousand eggs. All of these characteristics mean that these eating and breeding machines can expand into new areas in no time.

Consider what happened in Australia. In 1935 Reg Mungomery from the Queensland Bureau of Sugar Experiment Stations brought 101 cane toads from Hawaii to northern Queensland to breed them. Soon thereafter, scientists released 2,400 young toads into cane fields in an attempt to control the sugarcane beetles. But there was a problem: the beetles lived on the upper stalks of the cane, and the toads couldn't jump high enough to nab them. The toads would eat beetles that descended to the ground to lay their eggs but not enough to make a difference to the sugarcane yields. Furthermore, the toads fatally poisoned other animals—like monitor lizards—that ate the beetles. The toads found lots of other food and reproduced in their new environment. They marched down the coast of Queensland and westward toward the Northern Territory. By 2016 the toads occurred in more than 12.5 percent of the continent. Now, nearly ninety years after they were introduced, cane toads in Australia are estimated to be more than 200 million strong. And they are still advancing into new territory—at the rate of some thirty miles (fifty kilometers) per year.

But it isn't fair to view cane toads as merely an invasive species. In their native habitat in South America, cane toads are an integral part of the landscape and a crucial node in the food web, serving as both predators and prey. They eat invertebrates and small vertebrates and are eaten in turn by caimans and some birds and snakes. The toads' low, monotonous trill—reminiscent of a distant generator—is the perfect white noise to put a biologist to sleep after a long day in the field. Cane toads provide a good lesson for us not to introduce species into areas where they do not belong and to appreciate them where they do.

Soon after the streetlight turns off, our corpulent toad waddles into the nearby shrubbery. She will spend the day digesting her hearty meal, and as soon as the street-light comes back on, she will once again fill her belly with falling insects.

Common Midwife Toad

Alytes obstetricans

(EUROPE)

First light reveals a spectacle that challenges our perception. An olive-brown, warty frog peering up from the leaf litter has strings of eggs wound around his hind legs.

We are walking in an oak and beech woodland in western France on this humid morning. Like us, this frog has been up all night. He is now foraging for breakfast. Once the temperature warms and the surface of the ground begins to dry, our egg-carrying

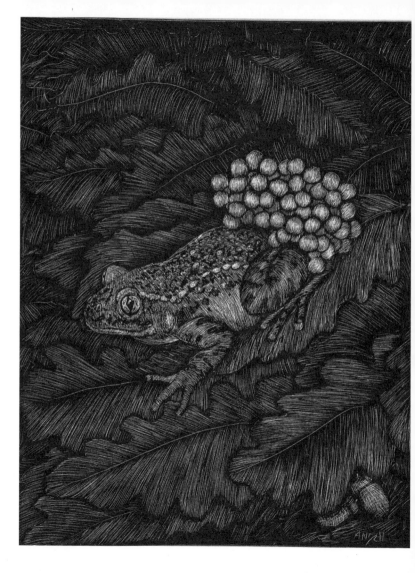

male will find a hiding place under a rock or log or disappear into a hole. Why? As a responsible dad, he won't risk having his eggs shrivel into tiny, hard beads. Fatherhood trumps foraging.

But why is he strolling around with eggs on his legs in the first place? After fertilizing his mate's eggs on land, a male common midwife toad thrusts his hind legs into the egg mass and wraps the strings of forty to fifty eggs around his ankles and hind legs. He then carries the eggs around for about a month, hanging out in damp places to keep them moist. When they're ready to hatch, he jumps into a pond, puddle, or water-filled ditch. And there the egg capsules burst open and the tadpoles develop until they metamorphose into froglets.

Imagine trying to walk with a string of twenty-five tennis balls tied around each leg.

This is what it must be like being this two-inch (fifty-one-millimeter) frog encumbered by dozens of eggs. Would you be slow and clumsy? Could you escape if a snake lunged toward you? Could you move fast enough to catch insects? There is a cost to parental care. The more eggs the males carry, the less distance they cover in a hop. And compared with their noncarrying counterparts, egg-laden males take longer to flip back over. That's a liability if a predator attacks.

Common midwife toads have a claim to fame in the annals of evolutionary history as the stars of the infamous midwife toad incident. During the early 1900s, Austrian biologist Paul Kammerer disagreed with the Darwinian theory of evolution by natural selection. Instead, he favored Jean-Baptiste Lamarck's theory that organisms pass to their offspring characteristics they acquire during their lifetimes. (As an example, Lamarck believed that by stretching to reach leaves high up in the trees, giraffes gradually passed on longer and longer necks to their offspring, resulting in our present-day long-necked giraffes.) Kammerer conducted elaborate experiments in the hope of supporting Lamarck's theory, one of which involved forcing common midwife toads to mate in water rather than on land as they normally do. In 1906 he published his results. He reported that the experimental male frogs had developed black "nuptial pads" on their thumbs. Males of many frog species that breed underwater have these pads, which are used for grasping mates. The common midwife toad is not such a species, but Kammerer also reported the presence of nuptial pads in the next generation of frogs he bred—evidence that they had inherited this acquired characteristic.

Gladwyn Kingsley Noble, curator of herpetology at the American Museum of Natural History, doubted the results. He examined the black patches microscopically and found that the areas had been injected with ink. Whether Kammerer falsified his data or a lab assistant sabotaged his experiment, we don't know. Either way, these frogs do not develop nuptial pads when they mate underwater, and Lamarck's theory remains discredited.

Our male has so many eggs wound around his legs that he must have mated with more than one female. In fact, a male common midwife toad can carry up to three clutches of eggs (about 150 in total) at once. This egg-carrying behavior might seem bizarre, but five other species of midwife toads have the same unusual paternal practice. Four of those species live in Europe, and the other one in Algeria and Morocco.

8 AM

Golden Poison Frog

Phyllobates terribilis

(SOUTH AMERICA)

The banana-yellow frog boldly foraging for insects in this dense, humid Colombian rainforest seems unafraid of us—for good reason.

Golden poison frogs, found only in Colombia, range in color from bright golden yellow to golden orange to pale metallic green. But they don't start life with those colors. Juveniles are black with golden-yellow racing stripes running down the sides of their bodies. By the time they are four or five months old, their stripes have disappeared and their black bodies have become bright and uniformly colored, warning even more strongly, "Don't eat me! I'm poisonous!" These frogs, active during the day, are one of the most toxic animals on Earth, so poisonous that they don't try to hide when disturbed. They simply hop away—and not particularly far at that.

The skin of golden poison frogs contains massive quantities of potent toxins that affect the nervous systems of almost all animals that ingest them. These toxins prevent nerves from transmitting impulses and lead to muscle paralysis, convulsions, and heart failure. Golden poison frogs are at least twenty times more toxic than their close relatives. A microscopic amount of the toxins would kill you if it entered your bloodstream through an open wound. One frog the size of a small banana pepper contains enough toxins to kill ten humans or twenty thousand mice. Considering there is no known antidote for its toxin, this is one formidable frog. In the scientific description of these frogs, the authors emphasized, "The new species is potentially dangerous to handle," and they advised wearing rubber gloves when doing so. No wonder the scientists named it *terribilis*.

Does a species this poisonous have any predators? Probably very few. One known predator is the fire-bellied snake (*Leimadophis epinephelus*), which can eat other poisonous frogs as well. Because fire-bellied snakes are small, they cannot swallow adult golden poison frogs and eat only juveniles. Humans might be the only other predator of these frogs. Some Indigenous peoples of northwestern Colombia use the toxins of golden poison frogs and two other poison frog species (also *Phyllobates*) to poison their blowgun darts. The three species have similar toxins, but golden poison frogs have much greater amounts. Emberá Chocó and Noanamá Chocó Indians collect toxins from the two less poisonous species by either impaling a frog through its mouth with a stick or holding a frog near a fire. The injury and stress cause the frog to release its secretions, into which the Indians rub their darts.

Golden poison frogs, by contrast, are so toxic that they don't need to be killed for their secretions. Instead, a person confines a few frogs in a closed basket. After allowing one to emerge from the basket, the person grasps the frog between two leaves and restrains it with a stick held underfoot. A dart is rubbed several times over the frog's back and then set aside, poisoned point up, in a green plantain leaf to dry. The person poisons two or three darts using one frog before letting it hop away—no doubt traumatized but otherwise unharmed. The poison is effective at killing monkeys and larger prey, including deer and even spectacled bears. The toxins decompose when the game is cooked, so the diners are not poisoned.

Golden poison frogs don't make these potent toxins themselves. Wild-caught frogs brought into captivity lose their toxicity, and the offspring of captive individuals are not poisonous. Why? A zoo's or lab's offering of insects differs from the frogs' natural diet, and golden poison frogs are what they eat. Their toxins come from the particular insects—most likely beetles—that they consume in nature. This frog-beetle dependency exemplifies the interconnectedness of living organisms. Imagine what would happen if these beetles disappeared. Without their beetle diet and therefore their toxicity, our bold golden poison frogs would be easy pickings for snakes, birds, and other predators.

You may be wondering how these frogs, capable of packing such a powerful punch, don't give themselves or their mates a heart attack. Scientists have recently discovered that they are immune to their own toxins thanks to a single gene mutation. Three cheers for evolution!

9 AM

Pacific Horned Frog
Ceratophrys stolzmanni

(SOUTH AMERICA)

Morning means it's safe to emerge from hiding. This juvenile Pacific horned frog, now hopping about in the leaf litter, is hungry.

Pacific horned frogs are nocturnal, and they spend daytime in their burrows. So why is our juvenile foraging at 9 AM? It is taking advantage of daylight hours to eat while its elders are hunkered underground.

If you want to know what any given frog species eats, you might start by measuring

the width of an individual's head. Pacific horned frogs have large, wide mouths nearly half the size of their bodies. These frogs generally sit in one place, waiting for prey to amble by and then pouncing. They will eat practically anything that fits in their mouths: slugs, insects, centipedes, snakes, rodents, and even other frogs—including their own kin.

There is good reason, then, for the juveniles' paradoxically timed meals: fear of their elders. By foraging during the daytime, they avoid cannibalistic adults. Laboratory experiments support this idea. Juveniles move significantly less in the presence of adults. They freeze and try not to call attention to themselves.

These frogs live in seasonally dry forests in coastal Ecuador and Peru, where a short rainy season dictates their activities. Adults breed after heavy rains, and tadpoles meta-morphose into froglets less than a month after fertilization. The juveniles have only about three months to eat before burrowing underground to wait out the dry season. They need to eat as much as they can while they can to ensure that one day they too become carnivorous cannibals terrorizing their own offspring.

You may be feeling sorry for Pacific horned froglets. After all, their families seem unusually dysfunctional. But these juveniles take after their parents and, in their short lives so far, have already exhibited aggressive traits. As tadpoles, they are cannibalistic and frequently eat their siblings. They grab a tadpole with their massive jaw muscles, rip into it, tear out a chunk, and eat the rest piecemeal. This predaceous behavior is most un-tadpole-like: tadpoles typically draw water into their mouths and then pass the water across a filtering system designed to trap suspended particles such as algae. The tadpoles of other species of horned frogs are likewise voracious predators. Recall the mud-nesting frogs we met a few hours ago at the ranch in Joaquín V. González. Once the tadpoles are freed from their mud nest and can swim about their newly formed pond, they must dodge the tadpoles of hungry Cranwell's horned frogs (*Ceratophrys cranwelli*).

Aggression seems to be a hallmark of the eight species of horned frogs, all of which are found in South America. Don't pick up our juvenile frog. It is likely to bite you, as I can vouch from personal experience. Horned frogs have two fang-like "teeth" (odontoids) on the bottom jaw, and when these sink into your finger, it hurts. In fact, Argentine folklore tells that a horse bitten by a horned frog will die. While studying Cranwell's horned frogs in Argentina, I met people who were deathly afraid of the frogs, mistakenly assuming that they were not only aggressive but also poisonous. Horned frogs seem to be fearless and will lunge at potential predators many times their own size. One day, my daughter's curious five-month-old kitten, Socks, approached a bowl containing a newly metamorphosed Cranwell's horned frog. The froglet raised high on its front legs, inflated its body to appear larger, opened its mouth, and expelled some of that air in a loud hiss. Socks jumped back in fright.

Take another look at our frog. You might recognize it from a different context. Horned frogs are often called "pacman frogs" because they resemble the large-mouthed dot-eating creature in the popular 1980 video game Pac-Man. Or, you might have seen horned frogs in pet stores. Their bizarre appearance and behavior has made them popular pets.

10 AM

Tomato Frog
Dyscophus antongilii

(MADAGASCAR)

A stroll through a backyard garden reveals an orange-red tomato look-alike resting near a patch of pink periwinkles bordering a shallow pond.

Tomato frogs live in rainforests, wooded coastal areas, and scrub forests, but they also adapt well to farmland and urban environments near drainage ditches or people's homes. Anywhere with a bit of standing water for breeding and some muddy or sandy ground for burrowing seems to be acceptable to a tomato frog. Male tomato frogs do not look quite ripe—their color is a less vivid yellow orange. Malagasy call this frog *sangongon*, a name that mimics the male frogs' rhythmic low-pitched calls.

These frogs are nocturnal, so why is our frog sitting in plain sight midmorning? She was startled, perhaps by us. Tomato frogs generally stay hidden during the day, but when disturbed by diurnal predators, they inflate their lungs with air to puff up their already stout bodies, making themselves appear larger and more intimidating. If their threatening size and orange-red or yellow-orange warning coloration do not deter a hungry predator, the frogs have a backup plan. They secrete a thick, mildly toxic mucus, which can irritate the eyes of some attackers, including dogs and cats. But this mucus isn't just irritating and foul tasting. It is also sticky—so sticky that it can glue a snake's jaws together temporarily, giving the frog time to escape. Or the mucus might glue a predator to the ground—for up to forty minutes for a small snake. After mouthing a tomato frog, a predator might find dry leaves or dirt stuck to the inside of its mouth. While it tries to clear the debris, the frog escapes. These defensive secretions are nearly five times stronger than rubber cement!

The biggest threat to tomato frogs isn't snakes, dogs, or cats. It's humans. Because of their beautiful coloration and cute, dumpy appearance, tomato frogs are highly prized in the pet trade. In the past, collectors would gather thousands of tomato frogs from the wild and ship them across the world to be kept as exotic pets, but it is now illegal to export these frogs from Madagascar. Thanks to efforts begun by the Baltimore Zoo in 1994 to breed the animals in captivity, there is no longer a need to collect frogs from the wild for the pet trade. But tomato frogs aren't home free. Humans are also threatening their surroundings. Tomato frogs currently have the near-threatened conservation status because of habitat deterioration, fragmentation, and deforestation. Unfortunately, the tomato frog's message of "Don't tangle with me lest you become entangled" does not protect its home.

Conservation biologists frequently use tomato frogs as a "flagship species"—iconic ambassadors to raise awareness of the need for species and habitat protection. What better animal could there be, alongside charismatic lemurs and chameleons, than a round, ripe, red tomato look-alike to encourage people—both on and off the island of Madagascar—to value the flora and fauna of this biological hotspot and to preserve what natural landscape remains?

In trying to take a photograph, we have further disturbed our frog, who has inflated her body. Now, even more than before, she resembles a juicy Big Boy tomato—one you might slice and top with buttery mozzarella and a drizzle of basil pesto vinaigrette. She does not intimidate us, but we are respectful and abandon the photo shoot.

Bornean Foot-Flagging Frog

Staurois guttatus

(ASIA)

Water rushing over boulders and a crashing waterfall create such a racket we can hardly hear ourselves think. A green and brown Bornean foot-flagging frog the size of a large garlic clove sits on a rock next to this fast-flowing stream, waving his flashy feet at another male nearby.

Male Bornean foot-flagging frogs need to

communicate in this noisy environment—to announce their territories to other males and to attract females for mating. They accomplish this with a two-part system: one part is acoustic, a rapid two-note call; the other is visual. When another frog is nearby, a male arches his body, sticks a hind foot outward into the air, rotates it along an arc, spreads his toes, and waves his foot. His feet are olive green with contrasting white or light-blue webbing between the toes, which end in white toe tips. He might add other visual displays, such as raising his feet, drumming his hind legs, waving his front legs, pushing himself upright, or opening his mouth. Males spend much of their time engaging both parts of this system. Early in the day, they rely primarily on calling, but once light levels increase—in the late morning hours such as this—they add more visual displays. An acoustic signal alerts nearby frogs to the caller's presence, and then the caller starts waving his feet. Humans wave to say hello or goodbye. These frogs wave to communicate "Hey, I'm over here, look at me. This is *my* territory." Or "Hey, I'm over here, look at me. I'm a fit male who could fertilize your eggs."

It's not just adult males who flag their flashy feet. Females do too, likely to defend foraging sites from other individuals: "The insects here are *mine!*" Even juvenile Bornean foot-flagging frogs wave their feet—just days after they metamorphose. As the froglets age, the webbing between their toes becomes brighter, changing from translucent gray to white or light blue, which makes their feet more conspicuous when flagged. But why do juveniles wave their feet? Perhaps for the same reason as females:

to maintain space between individuals and stake out foraging sites.

And it's not just adult males who call in this noisy environment. Females of very few frog species call, but the Bornean foot-flagging frog is among those few. Because they lack vocal sacs—the throat pouches that males fill with air when they call—female Bornean foot-flagging frogs call instead by expelling air with their mouths open. But why call at all? In short, aggression. Unreceptive females produce a series of high-pitched notes when males come within one foot (about thirty centimeters) of them. It's their way of saying, "I'm not interested. Back off." In response, the males stop advertising and either retreat or stay put.

That's how these frogs communicate with each other in their noisy environment, but how do they manage to climb the wet, slippery surfaces of rocks near waterfalls? The frogs are experts at adhering to rough surfaces in fast-flowing water for a couple reasons. For starters, their small body size makes it possible. (Imagine the large cane toad we met a few hours ago in Honolulu trying to do this!) Equally important, their rough skin grips the surface when the frogs press their bellies and toe pads against wet rocks.

At night, when most other frogs along this cascading rainforest stream are active, Bornean foot-flagging frogs sleep on vegetation overhanging the water or near the water's edge, restoring their energy for tomorrow's attention-getting displays. It's time to rest those tired feet.

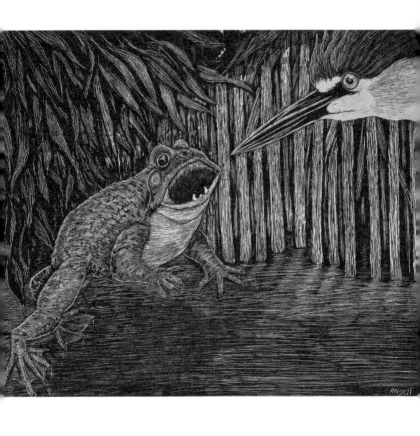

African Bullfrog

Pyxicephalus adspersus

(AFRICA)

The midday Kenyan sun, blisteringly hot, is causing a shallow pond to evaporate—and a father African bullfrog to bulldoze a path through the mud.

African bullfrogs are giant, at least as frogs go. The males of most frog species are smaller than their female counterparts, but African bullfrogs are not most frog species. These males are about twice as large as the females, reaching the size of dinner plates and weighing over 3 pounds (1.4 kilograms). Both males and females consume large prey:

frogs (including members of their own species), lizards, snakes, birds, and mammals. Fang-like odontoids in the lower jaw allow the frogs to hold onto their wriggling prey.

Males use their odontoids in other ways as well. Often called "pixies" in the popular lingo, the frogs live in savannas and semiarid to arid regions. Males gather in temporary ponds, which form after the first heavy summer rains, and belt out their "whoop, whoop" call that mimics lowing cattle. Smaller, younger males hang around the periphery hoping for a chance to mate, while larger and older males fight to defend their territories. Snorting, the frogs lunge toward each other with open mouths. They use their jaws to grab opponents by a leg, puncturing it with their odontoids, then hurl their rivals into the air. The frogs frequently wound each other during these territorial battles.

Male African bullfrogs are devoted fathers. After a pair lays and fertilizes about three to four thousand eggs in the water, the mom takes off. The dad stays and guards the eggs. Once the eggs hatch, the tadpoles swim around the pond in dense aggregations. The male stays with his offspring for at least three weeks and often until the tadpoles metamorphose into froglets some four weeks after hatching. Without sufficient rain, these ponds shrink in size and depth, and the tadpoles become stranded. Dad to the rescue! He digs a channel by flattening grasses and sedges and digging through the mud with his hind legs. These channels, ranging up to sixty feet (about eighteen meters), allow tadpoles to move into deeper, cooler water, where they have a better chance of metamorphosing before their pond becomes baked earth.

Fathers are fiercely protective of their offspring and will attack predators, such as water birds, that try to eat the eggs and tadpoles. Their odontoids and large size come in handy, but their guarding behavior comes at a cost. Defending males might be injured or killed, yet they seem to have no fear. As we approach this pugnacious and protective dad, he inflates his body, hisses, and lunges at us with mouth open. Only after that show of bravado does he turn and jump into the water. There is even a report of a male African bullfrog lunging at a lion that got a little too close to his offspring!

Once the rainy season ends, our African bullfrog will store water in his bladder and burrow into the soil to wait out the dry season. He will form a thick, multilayered cocoon of parchment-like shed skin, which will greatly reduce the amount of water his body loses. All but his nostrils will be covered, so he won't be able to open his eyes. His rate of metabolism will drop. He can stay underground for up to eight months, if the African dry season lasts that long.

Thanks to American cartoonist Gary Larson of *The Far Side* fame, African bullfrogs are part of popular culture. Larson had a pet pixie when he was a kid. He used pixies as an archetype for frogs in his cartoons, capturing their massive heads and wide mouths. My favorite is the comic with six frogs sitting on lily pads and one deflated frog rocketing into the air. One plump pixie exclaims, "Whoa! . . . Stuart blew his air sac!"

1PM
Rockhole Frog
Litoria meiriana

(AUSTRALIA)

A rockhole frog the size of a lima bean sits on a hot, dry rock in full sun. We're drenched with sweat as we stumble up this creek bed in the Australian Northern Territory. How can this frog sit in the hot sun?

Frogs typically cannot tolerate high temperatures. They lose water quickly through their permeable skin, and small frogs are especially vulnerable to drying out. So why doesn't our little lima bean overheat and shrivel into a crispy critter? Some animals use evaporative cooling to avoid lethally high body temperatures: Humans and horses sweat, and as the perspiration evaporates, it wicks away heat to cool the body. Dogs, sheep, and pigs pant, passing cooling air over their mouths and lungs. New World vultures and storks defecate on their legs; as the liquid evaporates, heat is drawn from the birds' bodies. But frogs don't sweat, pant, or poop to stay cool. Rockhole frogs cool themselves by evaporating—and then absorbing—water through their skin.

The key to this cooling process is the frogs' miniature size: they have a large amount of skin relative to the volume of their bodies. While they lose a lot of water quickly as a result, they also rehydrate quickly. To our left, a frog sits on the wet ground with his hind legs splayed. He is "drinking" water through his belly skin. After spending up to fifteen minutes on hot rocks, rockhole frogs either move onto wet soil or wet rocks to soak up moisture or they dive into the water and take a swim. And then it's back on the hot rocks again. They stay near water so that a pelvic or whole-body soak is never far away.

That's *how* these frogs can hang out on hot rocks during the day, but *why* do they do it? The frogs may benefit from a warmer body temperature, which allows for greater mobility, increased digestive function, and faster growth rates. Also, by being active during the day, their menu of insects differs from that of nocturnal frogs, and that means less competition for food. Another benefit is that they can forage when it's too hot for many of their predators, such as snakes, to be skulking about on the rocks.

What's in a name? The etymology of scientific names is supposed to be meaningful.

The International Code of Zoological Nomenclature allows silly names, but they should ideally be relevant, as in the names of three species of tiny frogs belonging to the genus *Mini*: *Mini mum*, *Mini scule*, and *Mini ature*, all found in Madagascar. Sometimes the specific epithet is chosen in honor of the person who collected the first recorded specimens—for example, *Rhinoderma darwinii* (Darwin's frog). Other times, it reflects some attribute of the species—for example, *Phyllobates terribilis* (golden poison frog). Still other times, we name species according to where they live. The specific name of our little frog, *meiriana*, hails from the local Aboriginal word in the Northern Territory for "rockhole." These frogs take shelter in holes in the rocks near streams and creeks.

The sudden movement we just detected in our peripheral vision is a rockhole frog hopping at high speed across the surface of the water, without sinking, just like a flat stone being skipped. Something must have frightened it. For the rockhole frog, it pays to be small and lightweight in more ways than one.

Giant Frog
Cyclorana australis

(AUSTRALIA)

When we meet a greenish-brown giant frog on this sunny afternoon in northern Queensland, she is basking in the warmth next to a woodland pond.

At about four inches (ten centimeters), giant frogs are only giant by Australian standards. They are half the size of the African bullfrog we met at noon. This female mated about a month ago, laying six thousand eggs in clusters that sank to the mud at the bottom of the pond. In another week, her tadpoles will metamorphose. Soon thereafter, she will be nowhere to be found. Giant frogs are the most encountered frogs in northern Australia—but only during the wet season, when you can often see them basking by temporary ponds during the day. Once the wet season ends, these ponds dry up and turn into cracked mud hollows. Little to no rain falls during the dry season.

Before the landscape dries up completely, while the sandy soil is still moist, our giant frog will dig a shallow burrow about five inches (thirteen centimeters) deep. And there she will estivate—the summer's equivalent of hibernation. For the first two to three months, she will absorb water from the soil in her burrow and store it in her bladder, increasing her body mass by 35 percent. Once the soil is so dry that her skin can no longer absorb water, she will spend the next forty-five days or so forming a cocoon.

This cocoon, composed of about thirty-four layers of skin cells, will cover everything but her nostrils and reduce water loss from her body, allowing her to stay cocooned underground for up to six months. After the first significant rains, she will break through her cocoon and emerge above ground. Light! Water! Food! The cycle will begin anew, with mating and egg laying during the wet season.

Giant frogs make tasty hors d'oeuvres for monitor lizards, which use their powerful claws to unearth the frogs from their shallow burrows. Would deeper burrows better protect these frogs from predators? Perhaps, but the shallow burrows enable the frogs to be more clued in to environmental conditions and thus more responsive to the beginning of the wet season. And, perhaps not coincidentally, the burrows are just deep enough to protect the frogs from low-intensity grass fires, which often sweep across the landscape during the dry season.

Giant frogs belong to a group of Australian species called "water-holding frogs," all of which depend on water that is stored in their bladders while they are cocooned underground. Sometimes during walkabout, Australian Aborigines dig up eastern water-holding frogs (*Cyclorana platycephala*), gently squeeze out a cup of water, and quench their thirst. They replace the frogs in their burrows and re-cover them with dirt out of respect, but unfortunately, the frogs are still left with their water tanks on empty. We'll never know the details of how people discovered this unusual source of drinking water, but it must have been a combination of human ingenuity and knowledge of the species' natural history.

Water-holding frogs, such as our giant frog, might have inspired an Australian Aboriginal creation story about a corpulent frog named Tiddalik, who satisfied his terrific thirst one morning by drinking all the fresh water from the billabongs, rivers, and lakes. After plants and animals began to dry up and die, wise old Wombat suggested that if someone could make Tiddalik laugh, he might expel the water. Animal after animal tried to make the bloated frog laugh, but none was successful, not even Kookaburra with his funniest jokes. Finally, Eel balanced on his tail, danced, and twisted his wormlike body into knots that he couldn't undo. Tiddalik's eyes lit up and laughter rumbled from deep within his belly. Water surged from his mouth and filled the billabongs, rivers, and lakes once more.

Tiddalik and giant frogs—fanciful and

bona fide water-holding frogs—capture our imagination and allow us to view the world through a different lens. Thanks to her cocoon and full bladder, our giant frog can rest while mammals and birds above ground search far and wide to quench their thirst during the long, hot dry season.

Yucatán Shovel-Headed Treefrog

Triprion petasatus

(MEXICO, CENTRAL AMERICA)

Midafternoon during the dry season is hot in this Mexican thorn forest. An olive-brown treefrog has found refuge in a tree hole, facing outward with its head plugging the entrance.

This isn't just any frog taking refuge in a tree hole. The Yucatán shovel-headed treefrog peering at us is specially adapted

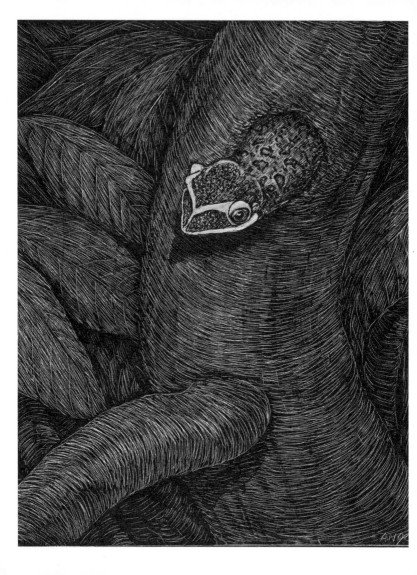

for such behavior. Its head, about one-third the length of its body, forms a bony casque. The skin on its head is completely attached to (coossified with) its skull bones. Its upper jaw bones are greatly expanded, forming a broad upturned shelf with a serrate edge, such that its snout extends far beyond the leading edge of its lower jaw. Add to that a bony knob sitting above each large, protruding eye, and the whole effect is of an otherworldly creature. Once plugged in the tree hole, a Yucatán shovel-headed treefrog is nearly impossible to remove. Its behavior protects it from predators, and its coossified casque head helps prevent it from drying out on this suffocatingly hot day. The treefrog's specific name—*petasatus*—is derived from Latin and means "with a hat on."

If you want to impress friends at your next dinner party, mention phragmosis—the technical term for using a body part to plug a hole or burrow to defend against predators. Several other species of casque-headed treefrogs also display this behavior, as do other animals. When threatened, some trapdoor spiders found in both the Old and New World biogeographical regions plug their burrow entrances with their abdomens, which end in hardened discs. In the Neotropics, worker turtle ants plug their nest entrance with their large, armored heads. And in Africa, worker door head ants have saucerlike shields on their heads, which they use to plug their nest entrance to deter predatory ants and other invertebrates from intruding.

If we were to return to this thorn forest after a heavy downpour during the wet season, our frog would not be passively plugging up a tree hole. It would be following the duck-like "quack-quack-quack" of like-minded neighbors from shrubs and low trees to water-filled depressions in rock outcrops and shallow limestone basins. There, female treefrogs would find males competing for their favor. They might even be simultaneously clasped by more than one male in a mating frenzy called an "explosive breeding event." Once paired, a female and her mate would deposit about fifteen hundred fertilized eggs in a shallow basin perhaps already occupied by the eggs of other Yucatán shovel-headed treefrogs. This basin will most likely not provide enough food for thousands of growing tadpoles. They will consequently turn into little predators, eating peers of their own and other species, just like the tadpoles of the Pacific horned frog we met this morning. It is good practice for later life, when as adults, they will eat small frogs.

But until that gully washer, the frogs must survive the dry season. As we inch closer to photograph our phragmotic frog, it opens its bulging, widely spaced, coppery eyes and stares at us. It knows it's safe—both from us and from the dry air.

4 PM

Oriental Fire-Bellied Toad

Bombina orientalis

(ASIA)

By this point in the day, we've met several species of poisonous frogs that advertise their toxicity with warning colors. Now we are meeting another—and this one adds a twist. A showy, warty, bright-green frog with jet-black spots hops near a Korean rice paddy. As we bend down for a closer look,

this Oriental fire-bellied toad treats us to a bizarre spectacle.

Our frog contorts its body into an astonishing posture called an "unken reflex." With its back arched and legs raised, it displays the fiery red and black of the undersurface of its feet, the sides of its belly, and its throat. We disturbed it, and it is warning us, "My skin glands brim with poisonous secretions." When an Oriental fire-bellied toad is in a resting position on the ground, its vivid underbody is concealed. It flashes the red and black colors only when predators venture too close. If the predators, including humans, ignore this warning, milky toxin-filled mucus will ooze out from the frog's skin and coat its body. These volatile secretions are a surefire deterrent: not only do they smell a bit like onions, but they will sting your nose and irritate your hands.

Oriental fire-bellied toads are enigmatic. For starters, the name of these striking creatures is a misnomer. Their backs are covered with small warts, but they are not toads in the strict sense. They are active mainly during the day, except during mating season, when they are most active at night. And they vary in color and behavior depending on where they live.

If we were to visit Jeju, the largest of South Korea's islands, we would see that the resident Oriental fire-bellied toads are dark brown and rather uniform instead of bright green with black spots. And their undersides are much less vivid than the frog in front of us. We would also see differences in the frogs' behaviors. Instead of moving about on the ground as our frog does, individuals on Jeju Island spend most of their time underwater, motionless under rocks or leaves.

That's not all. Island males spend less time calling than mainland males. Could these differences be related to the risk of predation? Clever investigators, hoping to find out, placed clay models of frogs on rocks at the edges of creeks in both locations. Twenty-four hours later, the scientists found bird beak and rodent incisor marks on forty-five of the six hundred models. Predators attacked the frog models on the island more than twice as often as they attacked those on the mainland. On the island, where the chance of getting eaten is higher, frogs keep a low profile.

People have used frogs in folk and traditional medicines for centuries. Now we know the scientific reason for the animals' healing effectiveness: antimicrobial peptides. These short chains of amino acids (the building blocks of proteins) defend against pathogens, including bacteria, viruses, fungi, and parasites. Diverse organisms rely on them as crucial components of their immune systems, but frog skin happens to be the most abundant natural source on Earth, with over one thousand unique antimicrobial peptides identified thus far. These peptides protect frogs from pathogens in their moist environments. Some, laboratory experiments have revealed, can fight human pathogens as well.

Could we one day exploit the complex cocktail of antimicrobial peptides in frog skin for therapeutic use in humans? Perhaps, and Oriental fire-bellied toads are helping us reach that goal. Bombinin peptides isolated from the defensive skin secretions of Oriental fire-bellied toads have antimicrobial effects against some bacteria, including strains of *Escherichia coli* that can cause severe stomach cramps and diarrhea, urinary tract infections, and pneumonia. Bombinin peptides can also fight the yeast *Candida albicans*, which is responsible for vaginal yeast infections, diaper rash, and thrush. Furthermore, bombinin peptides have potent anticancer effects on human hepatoma (liver cancer) cell lines. Perhaps one day, we'll be able to use synthetic bombinins to target and kill cancer cells without the severe side effects that many current drugs have on healthy cells.

Meanwhile, our little green frog with a fiery-red belly, secure in the knowledge that its unken reflex will scare off foes, continues to forage near the rice paddy.

Darwin's Frog
Rhinoderma darwinii

(SOUTH AMERICA)

A pudgy, emerald-green Darwin's frog soaks up the day's last rays of sunshine. He opens his mouth and out hops a tiny replica of himself.

Our frog has found a rare sunny spot in this temperate rainforest in southern Chile. Nearby on a patch of wet moss, another Darwin's frog chirps—"peeep, peeep, peeep." Our thoughts return to the sight we just witnessed, and we wonder, Had our pudgy frog tried to eat a baby? It took scientists nearly a century to answer that question.

It all began in 1834, when young Charles Darwin was exploring southern Chile during his voyage on the HMS *Beagle* and found an unusual-looking frog with a nose like Pinocchio's. He later found several more of these frogs near Valdivia, in central Chile. Darwin wrote in his journal that the frogs were "very pretty & curious" and sent specimens back to England. In 1841 two French scientists, André Marie Constant Duméril and Gabriel Bibron, named the species *Rhinoderma darwinii* (*rhinoderma* means "nose skin" in Greek). Seven years later, a French zoologist, Antoine Guichenot, examined some of Darwin's specimens and found tadpoles inside one of the frogs. Guichenot assumed the frog was female and that Darwin's frogs give birth to live young rather than lay eggs. But then, in 1872, Spanish zoologist Marcos Jiménez de la Espada found tadpoles in the vocal sacs of five other specimens. How the tadpoles had gotten into the vocal sacs remained a mystery for another sixty years.

These frogs, the size of small gherkin pickles and found only in Chile and Argentina, lay and fertilize five to fifteen eggs on land. The male stays near the clutch, and once the eggs are about to hatch, he slurps them into his mouth. They slide into his vocal sac, which will expand greatly as the tadpoles develop over the next two months. While brooding his young, a male does not call. Once they transform into little froglets, they crawl from their father's vocal sac back into his mouth. He then opens his mouth, and out hop the froglets, each about the size of your pinky fingernail.

Why is our male soaking up the day's last rays of sunshine? Sunbathing is unusual for frogs because their skin loses water easily.

Yet during my fieldwork with Darwin's frogs, I frequently found them sitting in sunny, wet spots. I was shivering in long underwear, a wool sweater and pants, a down jacket, and heavy-duty rainwear, so I shared their desire to seek sunshine. But I wanted to understand the behavior, so I measured the temperature where I found each frog—on the surface of dirt, moss, or fallen leaves. I determined that during the coldest part of the breeding season, brooding males hang out in warmer spots than do females and nonbrooding males. By heating their bodies this way, brooding males help the tadpoles develop faster. Best to cook those tadpoles as quickly as possible, for if a predator eats a brooding male, up to sixteen lives may be lost.

Darwin's frogs are the only extant frogs in the world that brood their young in their vocal sacs. Until a few decades ago, however, they had a close relative. Chile Darwin's frog (*Rhinoderma rufum*) once inhabited a small area in central Chile, but the last individual was seen in 1981. Male Chile Darwin's frogs brooded their tadpoles in their vocal sacs for about two weeks (instead of two months) and then burped them into water, where the young continued to develop until metamorphosing about three months later. Humans have destroyed the forest habitat of Chile Darwin's frogs, replacing native trees with nonnative pine trees, which likely explains the species' apparent extinction.

Sadly, Darwin's frogs also are disappearing. Their populations are small and isolated because of habitat destruction and a pathogenic fungus named *Batrachochytrium dendrobatidis*, or *Bd* for short. *Bd* was spreading around the world and killing frogs at least as early as the 1970s, but we didn't know it until 1999, when scientists identified and described the fungus. Conservation biologists now implicate *Bd* in the decline or extinction of at least two hundred species of frogs.

It would be tragic indeed to lose this unique frog. To safeguard against this possibility, Chilean scientists are breeding Darwin's frogs. Within the glass walls of terraria, males "peeep," females lay eggs, tadpoles incubate, and baby frogs hop out of their fathers' mouths. Scientists hope to one day release the captive-bred frogs into protected areas free of the killer fungus.

6 PM

Reticulated Humming Frog

Chiasmocleis royi

(SOUTH AMERICA)

Get ready to meet the ultimate odd couple. Dusk brings a hairy brown tarantula out of its burrow. It stations itself at the entrance. Soon its housemate—a reticulated humming frog—emerges.

We are in the Tambopata National Reserve in southeastern Peru. This preserved portion of Amazon rainforest is one of the world's most biodiverse areas, home to over 1,200 species of plants, 1,200 species

of butterflies, 600 species of birds, and 103 species of amphibians. And, of course, it is chock-full of complex interactions among the resident species. We are witnessing one such interaction: a tiny microhylid frog and a spider nearly four times its size living together in a burrow.

Earlier in the day, we met two other microhylid frogs—a Wandolleck's land frog transporting his froglets and a tomato frog warning, "Don't tangle with me!" The family Microhylidae consists of over 700 species found in the Americas, Asia, northern Australia, and Africa, including Madagascar. Many are small, hence the *micro* part of the name, and many have round bodies with pointed heads and narrow mouths, prompting the moniker "narrow-mouthed frogs." The spiders in this unusual housemate relationship belong to the family

Theraphosidae—predatory, hairy ground spiders often called tarantulas. This living arrangement between microhylids and theraphosids is widespread, having been reported in various species from the United States, Peru, Bolivia, Paraguay, India, and Sri Lanka. In some cases, home is a terrestrial burrow; in others, a tree hole.

The tarantulas shelter in their burrows for most of the day. Then, at dusk, they emerge and hang out near the entrance, waiting to ambush their prey—large insects, mice, even frogs. Like the tarantulas, the reticulated humming frogs stay in their burrows during the day, coming out at dusk to forage for ants nearby. Sometimes several reticulated humming frogs live together with a tarantula housemate. By daybreak, both tarantulas and frogs have returned to the safety of their shared abode.

When the frogs emerge from the burrows, they either push past the tarantulas or crawl under them. The tarantulas show no response, only occasionally shifting their legs or raising their bodies. Researchers have found that these tarantulas have a taste for frogs: the spiders readily ate five of the six species offered to them. The tarantulas pounced on individuals of the sixth species—túngara frogs, which we will meet later this evening—but released them unharmed. In contrast, the tarantulas ignore reticulated humming frogs, even when the frogs jump or crawl rapidly nearby. How do the spiders recognize their housemates? It isn't by sight—theraphosids have poor vision. It isn't by size, either—the tarantulas prey on other similarly sized frogs. Like many other spiders, they may rely on chemical cues. Túngara frogs have noxious skin secretions, which likely explains why the tarantulas did not eat them during the experiment. Reticulated humming frogs may also be protected thanks to toxic or otherwise unpalatable skin secretions.

Why forge such an unusual alliance? What's in it for both parties? The tiny frogs gain a humid but cool daytime refuge, protected from predators by their tarantula bodyguards. And the tarantulas? Ants eat spider eggs and spiderlings, and reticulated humming frogs eat ants. The tarantulas may then end up with more surviving offspring if they share their burrows. But this is only a hypothesis; field biologists would need to perform experiments to work out the exact nature of this relationship. Scientists thrive on such works in progress, and nature offers endless opportunities for those who love the thrill of discovery.

Rosenberg's Gladiator Frog

Hypsiboas rosenbergi

(CENTRAL AND SOUTH AMERICA)

As the sun sets and darkness settles in, a male Rosenberg's gladiator frog descends from the canopy of a Panamanian rainforest. His night starts with a moral question: Should he build a nest or steal one?

Female Rosenberg's gladiator frogs choose mates on the basis of their real estate. A male can come to own a nest in two ways: by constructing his own or by seizing someone else's. Nest construction requires no

less than thirty minutes—and often several hours—of work, depending on location. Once the male has settled on a slow-moving stream, he must form a basin in the mud by pivoting on his belly, pushing out with his front feet, and kicking backward with his hind feet. The depression will quickly fill with water, at which point he will remove debris with his snout and build high the ramparts. That's a lot of work. So, when he spots the perfect basin on his neighbor's property, he's tempted to steal. Of course, that male isn't going to roll over and give away his night's work: he will fight— sometimes to the death—to defend it.

Gladiator frogs, which are about the length of your middle finger, earned their name from their pugnacious behavior—and their weapons. On each thumb, males have a sharp, bony spine covered in a fleshy sheath.

When an intruder challenges a resident for his nest, they fight. The frogs jab their unsheathed spines at their opponent's eyes and eardrums. These violent fights involve wrestling and blocking and are accompanied by growls and hisses. Males rarely emerge unscathed.

Once a male owns a nest, he must keep it in good condition. This requires nightly renovation. He uses his feet to push mud onto eroded ramparts and his snout to remove leaves and twigs that have fallen into the basin. From inside his nest, he calls to attract a female—"tonk-tonk-tonk," like someone beating on a solid piece of wood. An interested female will jump into the basin and investigate. If she finds the nest agreeable, she rubs the male's head and body with her hands, chest, and chin to signal her willingness to mate. At this point, with the clasping male on her back, she makes any final modifications, deepening and widening the nest to her liking and removing any remaining debris. By early morning, the pair has laid and fertilized two to three thousand eggs, now floating on the water's surface.

Once the female departs, the male's parental duties begin. He must guard his eggs from rival males attempting to steal his nest. If an intruder dips so much as a toe into the nest, the surface film will be disturbed and the eggs will sink to the bottom and die. The tadpoles hatch about forty hours after fertilization. Dad is free to go, but his offspring remain in the basin until rain washes them into the adjacent stream. There they complete development through metamorphosis.

Life doesn't have to be this hard, though. Parental care, from nest building to nest guarding, consumes a lot of time and energy, and fighting is dangerous. In places where there are many natural water-filled depressions, such as cattle hoofprints, or where the density of gladiator frogs is low, these time- and energy-consuming activities can be scrapped. In those places, a male generally calls by a puddle to attract a female, then once the eggs are fertilized, both parents abandon the clutch. Why waste resources when there are no intruders lurking about, attempting to steal what is not theirs?

The common denominator in the reproductive activity of Rosenberg's gladiator frogs, whether thieving, building, or just visiting, is their descent from the canopy—their daytime refuge—at dusk. Right about now, these agile treefrogs are taking long jumps down, covering a lot of distance quickly. There's never any time to waste.

Our male has chosen the high road and is constructing a basin in the squishy mud beneath a flowering lobster claw plant (*Heliconia*). We wish him luck attracting a mate and fighting off rivals.

Matang Narrow-Mouthed Frog
Microhyla nepenthicola

(BORNEO)

In the falling darkness, we watch two tiny Matang narrow-mouthed frogs perched at the edge of a green and red pitcher plant. Clasped in a mating embrace, they soon disappear into the pitcher. Five minutes later, the frogs have not yet emerged. Are they doomed?

We're walking along the edge of a lush rainforest in the Matang Range, Sarawak, and this is the first pair of these frogs we've seen.

It's no wonder. Matang narrow-mouthed frogs are one of the world's smallest frog species, slightly smaller than shelled almonds. But we can certainly hear them. It is now dark enough that males have reached their peak calling activity, and we have tracked their harsh, raspy notes to a patch of narrow-lid pitcher plants (*Nepenthes ampullaria*).

It turns out our initial worries were unfounded; our frogs purposely landed in the pitcher plant, where they will deposit and fertilize eggs on the inside wall. Once the eggs hatch, the tadpoles will develop in the pitcher's liquid. The tadpoles of this species are tiny, yet almost the size of adult males. Seventy percent of their total length is tail. After about two weeks in a pitcher, surviving on nothing but their own egg yolk, the tadpoles metamorphose into tiny froglets.

Why doesn't the plant digest the tadpoles? After all, pitcher plants are typically carnivorous. Narrow-lid pitcher plants, however, thrive on a different source of nutrients than their closest relatives: rather than catching bugs, they digest the leaves (and the occasional invertebrates) that fall into their pitchers. Compared to other pitcher plants, these have broad mouths and not much of a lid; they lack the wax-covered area that would cause prey to lose their foothold and slip to their deaths, and their digestive juices are less acidic. Narrow-lid pitcher plants are one of the smallest species in the genus, with pitchers less than four inches (ten centimeters) high. Thanks to the slow but steady trickle of nutrients from fallen leaves, individual pitchers can last more than six months. That's six times longer than the more carnivorous species. These plants often form a densely packed carpet on the ground—a perfect condominium of nurseries, not just for our frogs but also for over a dozen other animals, including mosquito larvae.

Biologists call Matang narrow-mouthed frogs "miniaturized" because they are smaller than four-fifths of an inch (twenty millimeters). The world's smallest vertebrate, *Paedophryne amauensis*, is in the same family. You'll find that distant cousin in Papua New Guinea—or you'll *try* to find it, as it's about the size of a pea. Most miniaturized frogs, including Matang narrow-mouthed frogs, live in the leaf litter of tropical wet forests, where they can stay moist to avoid water loss.

Does it pay to be miniature? Absolutely. Tiny frogs exploit the nooks and crannies in their environments that larger frogs cannot, including inside pitcher plants. They're safe from many vertebrate predators that don't bother eating tiny prey—why spend the time and energy for so few calories? And tiny frogs can eat different foods, such as mites and springtails, that larger frogs ignore. In doing so, they avoid competition with neighbors.

In this dense patch of available tadpole nurseries, how did our pair of Matang narrow-mouthed frogs choose their pitcher? Pitchers that are not too old, sufficiently filled with fluid, and nestled under the trees' shade are just right.

Túngara Frog

Engystomops pustulosus

(CENTRAL AND SOUTH AMERICA,
THE CARIBBEAN)

Floating on the surface of a roadside puddle, a male túngara frog calls to attract a female on this dark and rainy night. His hind legs are splayed, and his enormous, inflated vocal sac resembles a fully blown-up balloon.

We're in Panama, walking along a dirt road bordering a forest. We have timed our visit just right. The calls of túngara frogs begin around dusk and reach their peak sometime between 8 and 10 PM. An individual male, the size of a jumbo kalamata olive, might call more than five thousand times in a single evening. Our frog's call is a down-sweeping harmonic element—a whine. Alone in his shallow puddle, he is free of competition; any female that approaches and finds him worthy will be his.

Farther down the road, a cacophony of calls leads us to a water-filled ditch containing ten túngara frogs, each whining like the first but now adding several staccato "chuck" notes—"W-h-i-n-e, chuck-chuck-chuck." This complex call gives these frogs their onomatopoetic name: *tún* is the whine, and *ga-ra* are two chucks. The contrasting sounds send different messages. The whine attracts females, while the chucks provide them with additional information, such as the body size of the caller. Larger males, who utter lower-frequency chucks, fertilize a greater proportion of a female's eggs. As you might expect, calls that include chucks are more attractive to females.

Since females prefer complex calls, you might think that all males—whether alone or in a chorus—would include chucks. But complex calls are risky. In advertising their whereabouts to females, males also reveal themselves to frog-eating fringe-lipped bats (*Trachops cirrhosis*). Unfortunately for the frogs, these eavesdropping bats find it easier to locate the complex calls. So, the frogs must balance the costs and benefits of their calling behavior. If solo, like the puddle-based frog we first met, males should not risk becoming prey by adding chucks. But there is safety in numbers; it's okay to show off in a group. After all, if a bat approaches, it might eat someone else. If we were to visit the nearby island of Taboga, where there are no fringe-lipped bats, we would find that even solitary male túngara frogs embellish their calls with chucks.

As if frog-eating bats weren't enough of a distraction from mating, frog-biting midges (a species of *Corethrella* in the order Diptera) also flock to calling male túngara frogs. Like the bats, the midges can locate chucking males more easily than merely whining ones.

Female midges land on a frog's back and then walk up to its nostrils, where they help themselves to a blood meal. These miniature vampires need blood to produce eggs and are therefore determined—just like the pesky female mosquitoes that buzz around our ears—to find a host. Túngara frogs swat the midges with their front feet. Why don't the midges feed from a frog's back to eat in peace? The blood vessels are larger and more abundant around a frog's nostrils than on its back. The midges tolerate the swatting because nose blood is better.

Túngara frogs that live in the town up the road give even more complex calls than our forest frogs. Why? The streetlights in town drive away the eavesdropping frog-eating bats and frog-biting midges, so the town frogs can boast all they want with less risk of getting eaten or losing blood.

Back at our forest site, we've just frightened the frogs in the water-filled ditch, and they react as though we are bats. Each has stopped calling, deflated its balloon-like vocal sac, and swum away. If we stand quietly with our headlamps turned off, they will emerge from hiding and resume their whines and seductive chucks.

10 PM
African Clawed Frog
Xenopus laevis

(AFRICA, INTRODUCED ELSEWHERE)

Our headlamp beams reveal a frenzy of activity in a water-filled ditch running along a red-earth road in Uganda. An African clawed frog is preparing a large tadpole for dinner.

African clawed frogs, nocturnal and fully aquatic, don't have tongues. Whereas most frogs roll out their long, sticky tongues to hit their prey, then roll their tongues back in to swallow their meal whole, African clawed

143

frogs take food preparation to another level. As ambush predators, they remain still until prey—such as fish, insects, or other frogs—move close to their mouths. Then they open their mouths, generating rapid suction so that they can grab the victim with their teeth. They use the black claws on the three inner toes of their hind feet to shred large prey, then shove the pieces into their mouths with their front feet. These frogs eat just about anything they can catch, and they scavenge dead birds and mammals they find in the water. They are cannibals and will eat their own tadpoles and even fellow adults.

These pear-sized frogs live in just about any body of water—natural or human made, from aquatic oases in the desert to ice-covered lakes at high altitudes. They are also remarkably tolerant of diverse water quality, from stagnant ponds to clear, fast-flowing streams. During a prolonged drought, they burrow backward into the drying mud and hang out until rain comes. They can live a year without food. And like our mud-nesting frog tadpoles in their mud nests, African clawed frogs in their underground shelters switch from excreting highly toxic ammonia to less toxic urea so as not to foul their temporary dwellings.

Their ability to eat just about anything and live just about anywhere suggests that African clawed frogs released into a nonnative habitat are likely to survive. And that is exactly what they've done. Between the 1930s and 1960s, physicians used the frogs for a procedure called the Hogben test. A technician injected a small amount of a woman's urine under the skin of a female African clawed frog. If the frog laid eggs within twelve hours, the woman was pregnant. How did this pregnancy test work? Hormones in a pregnant woman's urine triggered the frog's egg release. The test was reliable, with an accuracy rate of 99 percent. But once researchers developed more advanced tests and the frogs were no longer needed, many found themselves discarded in local ponds and swamps. And, of course, they survived. African clawed frogs are popular in the pet trade because they are so hardy, undemanding, and, let's face it, weird looking, with their flattened bodies and small round eyes perched atop their heads. But when people tire of their pet frogs, they too dump them in local ponds. African clawed frogs are now considered an invasive species in the United States, Europe, Mexico, Chile, and Japan, where they often outcompete or devour native species. They also transmit parasites and diseases to native frogs.

Because African clawed frogs are so adaptable, we can easily breed and maintain them in aquaria. And that makes them excellent animals for experimental research. For decades, scientists have used African clawed frogs as subjects for developmental biology and biomedical research. In 1958 in work that would later win him the Nobel Prize, British developmental biologist Sir John Gurdon gave African clawed frogs the honor of being the first vertebrate to be cloned. In 1992 four female African clawed frogs received another distinction—they became amphibian astronauts, rocketing off the launch pad at the Kennedy Space Center in Florida as part of the laboratory research mission of the space shuttle *Endeavour*. The goal? To determine whether embryonic development would occur normally in the absence of gravity. Once in orbit, researchers injected the frogs with a hormone that made them lay eggs and then fertilized those eggs with sperm collected from male African clawed frogs. The cells began to divide, and four days later, the eggs hatched. Several days after that, *Endeavour* brought the tadpoles back to Earth, where they eventually metamorphosed into frogs. Although conceived in the cosmos, the frogs later produced normal offspring. African clawed frogs showed the world that the eggs of a vertebrate can develop from fertilization to a free-living stage in the absence of gravity. Mission accomplished.

Our frog has finished its tadpole dinner. It is now submerged in the murky water, with only its small, beady eyes and nostrils exposed. Its next meal might be a sibling.

Common Toad
Bufo bufo

(EUROPE, ASIA, AFRICA)

Slow down and swerve to the right! A common toad is crossing the road on this drizzly spring night in southern England. Why? To get to the other side—and to a breeding pond—of course.

Common toads spend most of the year living solitary lives under logs, rocks, or leaf litter in forests, fields, and pastures. But once temperatures warm in springtime, it is time to socialize. Common toads often migrate long distances (up to three miles, or five kilometers) to breed in their ances-

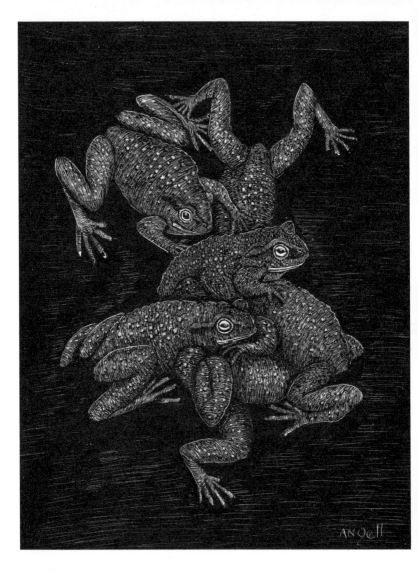

tral pond. Sometimes, males arrive at the pond before females; other times, males and females migrate en masse, on the same wet nights.

Once they arrive, it's a free-for-all—an explosive breeding event. Unmated males that encounter a mated pair climb aboard; the first male, in turn, kicks his hind legs in an attempt to dislodge his rival. Sometimes as many as a dozen males—maybe even more—surround the female in a "toad ball." Male common toads have poor powers of sex discrimination and will clasp just about any moving object they can wrap their front legs around: other males, other species of frogs, fish, vegetation, drowned females, even human thumbs. During this melee, females somehow festoon two long strings of up to six thousand eggs around aquatic vegetation as one or more males fertilize the black-and-white pearls. Then, in a matter of days—two weeks at most—mating is over. Time to return to the forests, fields, and pastures.

But not all common toads make it to the breeding pond and not all make it home. A toad will take a few hops on the road and then stop. A few more and then stop. This leisurely journey is risky. Cars squash many hundreds of thousands of toads every year during their breeding migration. In the United Kingdom alone, an estimated twenty tons of common toads are killed on roads every year. Common toads got their name because they are one of the most success-ful species of frogs on Earth; they are both abundant and widespread, appearing throughout most of Europe, central Asia, and northern Africa. But how much longer will common toads be common?

Humans have caused this road-crossing

mortality problem, but humans have also come to the rescue. When word goes out that toads are on the move, volunteers grab their rubber boots, thermoses of hot tea or coffee, and snacks and mount "toad patrols" at scattered sites across western Europe. As toads arrive at the edge of roads, volunteers scoop them up, place them in buckets, and release them on the other side. In Britain alone, there are several hundred registered "toad crossings"—places where volunteers spend their evenings ferrying toads across roads. Froglife, a UK national wildlife charity, sponsors a program called Toads on Roads. In 2021 this toad-shuttling service had over one hundred patrolling sites, where toad patrollers helped over eighty thousand toads cross roads. Sadly, patrollers still recorded over six thousand toad deaths at their sites.

People share the roads with common

toads in other ways as well. In many places, toad-crossing signs warn motorists of migrating toads. Roads that cross heavy migration routes are temporarily closed. Engineers build tunnels beneath and drift fences along roads. When the toads find that they can't cross the upright structures made of metal or plastic, they move along the fence until they come to an underpass—an escape route that delivers them to the other side.

It's impressive that this toad has convinced so many humans to help, considering our long, negative history together. During the Middle Ages, many Europeans associated toads with witchcraft. As familiars, these demons might carry out the witches' work or become key components in their magic potions and flying ointments.

If only these toads had the magical powers needed to escape our threats. Industry, urbanization, and other human forces have destroyed considerable toad habitat. It is only right and fitting that humans do all we can to help. And so, we pick up our toad and place it, facing the direction of deep, distant croaks, on the other side of the road. This is the end of our day but not the toad's journey.

Epilogue

With over 7,600 species of frogs worldwide, I could have chosen many different combinations of captivating animals for our field trip. We could have met a pebble toad on a Venezuelan tepui, rolling downhill like a small dislodged stone to escape from a tarantula. A strawberry poison frog depositing unfertilized eggs for her tadpoles to eat. A male hip-pocket frog carrying tadpoles in pouches along the sides of his body. An Izecksohn's Brazilian treefrog nibbling fruit and slurping nectar; as it moves from flower to flower, it just might pollinate the plant—a first for frogs! Not only that, but it can also help distribute the plant's seeds, which remain viable in its poop, throughout the forest—another first for frogs! A paradoxical frog tadpole that is four times larger than the frog into which it will metamorphose. A male holy cross toad so small relative to his mate that his front legs are too short to grasp her; instead, he attaches to her back with a sticky secretion. We'll just have to meet these—and many other frogs just as intriguing—on another twenty-four-hour field trip around the world.

I hope we get that chance. Biodiversity is declining worldwide. The International Union for Conservation of Nature (IUCN) reports that amphibians are the most threatened group of vertebrates, with an estimated 41 percent of species threatened with extinction. Compare that to 21 percent of reptiles, 13 percent of birds, and 27 percent of mammals. As we've seen, some of the species we met on our field trip are declining because of human-led habitat destruction or modification. Others are declining because of disease, overexploitation, pollution, introduced species, or climate change. The good news is that scientists and the general public are becoming more aware of the issues and are working on many different fronts to turn back the tide of disappearing amphibians. Conservation begins with passion, including from people willing to spend a day with frogs. Thank you.

Acknowledgments

I thank Joseph Calamia for his enthusiasm, support, and guidance during the writing of *Frog Day*. Joe reviewed all twenty-four chapters and offered helpful suggestions. His *oohs*, *ahhs*, and *wows* of amazement validated just how special and unique the frogs encountered on this field trip are. Several anonymous reviewers also provided helpful feedback on the manuscript. I thank Lily Sadowsky and the rest of the team at the University of Chicago Press for their efforts in bringing this project to completion. I am grateful to artist Tony Angell, who gave the frogs life through his beautiful drawings. Finally, thanks go to the many scientists who spent long hours in the field and in the laboratory studying the frogs highlighted herein. Their careful observations and creative experiments are critical for our understanding of the species and therefore for their protection.

Further Reading

Beebee, T. J. C. 1996. *Ecology and Conservation of Amphibians.* London: Chapman and Hall.

Bickford, D. P. 2001. "The Ecology and Evolution of Parental Care in the Microhylid Frogs of New Guinea." PhD diss., University of Miami.

Cane Toads in Oz: A Gigantic Frog in a Strange New Land. www.canetoadsinoz.com.

Crump, M. 2000. *In Search of the Golden Frog.* Chicago: University of Chicago Press.

Crump, M. 2013. *The Mystery of Darwin's Frog.* Honesdale, PA: Boyds Mills Press.

Crump, M. 2015. *Eye of Newt and Toe of Frog, Adder's Fork and Lizard's Leg: The Lore and Mythology of Amphibians and Reptiles.* Chicago: University of Chicago Press.

Crump, M. 2018. *A Year with Nature: An Almanac.* Chicago: University of Chicago Press.

Duellman, W. E. 2015. *Marsupial Frogs: Gastrotheca and Allied Genera.* Baltimore, MD: Johns Hopkins University Press.

Froglife. "Toads on Roads." https://www.froglife.org/what-we-do/toads-on-roads/.

Halliday, T. 1980. *Sexual Strategy*. Chicago: University of Chicago Press.

Lannoo, M. J., and R. M. Stiles. 2020. *The Call of the Crawfish Frog*. New York: CRC Press.

Ryan, M. J. 1985. *The Túngara Frog: A Study in Sexual Selection and Communication*. Chicago: University of Chicago Press.

Shine, R. 2018. *Cane Toad Wars*. Oakland: University of California Press.

Wahab, M. P. B. 2012. "Pitcher Plant (*Nepenthes ampullaria*) Choice by Frogs of the *Microhyla nepenthicola* and *M. borneensis* Complex for Breeding at Kubah National Park, Sarawak." Senior honors thesis, Universiti Malaysia Sarawak.

RELEVANT PEER-REVIEWED PAPERS

MIDNIGHT: WALLACE'S FLYING FROG

Emerson, S. B., and M. A. R. Koehl. 1990. "The Interaction of Behavioral and Morphological Change in the Evolution of a Novel Locomotor Type: 'Flying' Frogs." *Evolution* 44 (8): 1931–46.

Stückler, S., S. Cloer, W. Hödl, and D. Preinger. 2022. "Carotenoid Intake during Early Life Mediates Ontogenetic Colour Shifts and Dynamic Colour Change during Adulthood." *Animal Behaviour* 187:121–35.

1AM: HORNED MARSUPIAL FROG

Gagliardo, R., E. Griffith, R. Hill, H. Ross, J. R. Mendelson III, E. Timpe, and B. Wilson. 2010. "Observations on the Captive Reproduction of the Horned Marsupial Frog *Gastrotheca cornuta* (Boulenger 1898)." *Herpetological Review* 41 (1): 52–58.

Warne, R. W., and A. Catenazzi. 2016. "Pouch Brooding Marsupial Frogs Transfer Nutrients to Developing Embryos." *Biology Letters* 12 (10): 20160673.

2AM: MUD-NESTING FROG

Crump, M. L. 1995. "*Leptodactylus bufonius* (NCN). Reproduction." *Herpetological Review* 26 (2): 97–98.

Faggioni, G., F. Souza, M. Uetanabaro, P. Landgref-Filho, J. Furman, and C. Prado. 2017. "Reproductive Biology of the Nest Building Vizcacheras Frog *Leptodactylus bufonius* (Amphibia, Anura, Leptodactylidae), Including a Description of Unusual Courtship Behaviour." *Herpetological Journal* 27:73–80.

Philibosian, R., R. Ruibal, V. H. Shoemaker, and L. McClanahan. 1974. "Nesting Behavior and Early Larval Life of the Frog *Leptodactylus bufonius*." *Herpetologica* 30:381–86.

Reading, C. J., and G. M. Jofré. 2003. "Reproduction in the Nest Building Vizcacheras Frog *Leptodactylus bufonius* in Central Argentina." *Amphibia-Reptilia* 24:415–27.

Shoemaker, V. H., and L. L. McClanahan. 1973. "Nitrogen Excretion in the Larvae of a Land-Nesting Frog (*Leptodactylus bufonius*)." *Comparative Biochemistry and Physiology* 44A:1149–56.

3AM: WOOD FROG

Costanzo, J. P. 2019. "Overwintering Adaptations and Extreme Freeze Tolerance in a Subarctic Population of the Wood Frog, *Rana sylvatica*." *Journal of Comparative Physiology B* 189 (1): 1–15.

Gibbs, J. P., and A. R. Breisch. 2001. "Climate Warming and Calling Phenology of Frogs near Ithaca, New York, 1900–1999." *Conservation Biology* 15 (4): 1175–78.

Larsen, A. S., J. H. Schmidt, H. Stapleton, H. Kristenson, D. Betchkal, and M. F. McKenna. 2021. "Monitoring the Phenology of the Wood Frog Breeding Season Using Bioacoustics Methods." *Ecological Indicators* 131:108142. https://doi.org/10.1016/j.ecolind.2021.108142.

4AM: WANDOLLECK'S LAND FROG

Bickford, D. 2002. "Male Parenting of New Guinea Froglets." *Nature* 418 (6898): 601–2.

Diesel, R., G. Bäurle, and P. Vogel. 1995. "Cave Breeding and Froglet Transport: A Novel Pattern of Anuran Brood Care in the Jamaican Frog, *Eleutherodactylus cundalli*." *Copeia* 1995:354–60.

5AM: CRAWFISH FROG

Heemeyer, J. L., and M. J. Lannoo. 2012. "Breeding Migrations in Crawfish Frogs (*Lithobates areolatus*): Long-Distance Movements, Burrow Philopatry, and Mortality in a Near-Threatened Species." *Copeia* 2012 (3): 440–50.

6AM: CANE TOAD

Mittan-Moreau, C. S., C. Kelehear, L. F. Toledo, J. Bacon, J. M. Guayasamin, A. Snyder, and K. R. Zamudio. 2022. "Cryptic Lineages and Standing Genetic Variation across Independent Cane Toad Introductions." *Molecular Ecology* 31 (24): 6440–56. https://doi.org/10.1111/mec.16713.

7AM: COMMON MIDWIFE TOAD

Lange, L., L. Bégué, F. Brischoux, and O. Lourdais. 2021. "The Costs of Being a Good Dad: Egg-Carrying and Clutch Size Impair Locomotor Performance in Male Midwife Toads, *Alytes obstetricans*." *Biological Journal of the Linnean Society* 132 (2): 270–82.

Márquez, R. 1992. "Terrestrial Paternal Care and Short Breeding Seasons: Reproductive Phenology of the Midwife Toads *Alytes obstetricans* and *A. cisternasii*." *Ecography* 15:279–88.

8AM: GOLDEN POISON FROG

Myers, C. W., J. W. Daly, and B. Malkin. 1978. "A Dangerously Toxic New Frog (*Phyllobates*) Used by Emberá Indians of Western Colombia, with Discussion of Blowgun Fabrication and Dart Poisoning." *Bulletin of the American Museum of Natural History* 161:307–66.

Wang, S.-Y., and G. K. Wang. 2017. "Single Rat Muscle Na+ Channel Mutation Confers Batrachotoxin Autoresistance Found in Poison-Dart Frog *Phyllobates terribilis*." *Proceedings of the Natural Academy of Sciences* 114 (39): 10491–96.

9AM: PACIFIC HORNED FROG

Ortiz, D. A., D. Almeida-Reinoso, and L. A. Coloma. 2013. "Notes on Husbandry, Reproduction and Development in the Pacific Horned Frog *Ceratophrys stolzmanni* (Anura: Ceratophryidae) with Comments on Its Amplexus." *International Zoo Yearbook* 47:151–62.

Székely, D., D. Cogălniceanu, P. Székely, and M. Denoël. 2020. "Adult-Juvenile Interactions and Temporal Niche Partitioning between Life-Stages in a Tropical Amphibian." *PLoS ONE* 15 (9): e0238949. https://doi.org/10.1371/journal.pone.0238949.

10AM: TOMATO FROG

Evans, C. M., and E. D. Brodie Jr. 1994. "Adhesive Strength of Amphibian Skin Secretions." *Journal of Herpetology* 28 (4): 499–502.

11AM: BORNEAN FOOT-FLAGGING FROG

Endlein, T., W. J. P. Barnes, D. S. Samuel, N. A. Crawford, A. B. Biaw, and U. Grafe. 2013. "Sticking under Wet Conditions: The Remarkable Attachment Abilities of the Torrent Frog, *Staurois guttatus*." *PLoS ONE* 8 (9). https://doi.org/10.1371/journal.pone.0073810.

Grafe, T. U., and T. C. Wanger. 2007. "Multimodal Signaling

in Male and Female Foot-Flagging Frogs *Staurois gutta-tus* (Ranidae): An Alerting Function of Calling." *Ethology* 113:772–81.

Stangel, J., D. Preininger, M. Sztatecsny, and W. Hödl. 2015. "Ontogenetic Change of Signal Brightness in the Foot-Flagging Frog Species *Staurois parvus* and *Staurois guttatus*." *Herpetologica* 71 (1): 1–7.

NOON: AFRICAN BULLFROG

Balinsky, B. I., and J. B. Balinsky. 1954. "On the Breeding Habits of the South African Bullfrog, *Pyxicephalus adspersus*." *South African Journal of Science* 51:55–58.

Cook, C. L., J. W. H. Ferguson, and S. R. Telford. 2001. "Adaptive Male Parental Care in the Giant Bullfrog, *Pyxicephalus adspersus*." *Journal of Herpetology* 35:310–15.

Loveridge, J. P., and P. C. Withers. 1981. "Metabolism and Water Balance of Active and Cocooned African Bullfrogs *Pyxicephalus adspersus*." *Physiological Zoology* 54 (2): 203–14.

1PM: ROCKHOLE FROG

Tracy, C. R., K. A. Christian, N. Burnip, B. J. Austin, A. Cornall, S. Iglesias, S. J. Reynolds, T. Tixier, and C. Le Noëne. 2013. "Thermal and Hydric Implications of Diur-

nal Activity by a Small Tropical Frog during the Dry
Season." *Austral Ecology* 38 (4): 476–83.

2PM: GIANT FROG

Tracy, C. R., S. J. Reynolds, L. McArthur, C. R. Tracy, and K.
A. Christian. 2007. "Ecology of Aestivation in a Cocoon-
Forming Frog, *Cyclorana australis* (Hylidae)." *Copeia*
2007 (4): 901–12.

Withers, P. C., and G. G. Thompson. 2000. "Cocoon For-
mation and Metabolic Depression by the Aestivating
Hylid Frogs *Cyclorana australis* and *Cyclorana cultripes*
(Amphibia: Hylidae)." *Journal of the Royal Society of
Western Australia* 83:39–40.

3PM: YUCATÁN SHOVEL-HEADED TREEFROG

Duellman, W. E., and L. T. Klaas. 1964. "The Biology of the
Hylid Frog *Triprion petasatus*." *Copeia* 1964 (2): 308–21.

Lee, J. C., and M. L. Crump. 1981. "Morphological Correlates
of Male Mating Success in *Triprion petasatus* and *Hyla
marmorata* (Anura: Hylidae)." *Oecologia* 50 (2): 153–57.

4PM: ORIENTAL FIRE-BELLIED TOAD

Kang, C., T. N. Sherratt, Y. E. Kim, Y. Shin, J. Moon, U. Song,
J. Y. Kang, K. Kim, and Y. Jang. 2017. "Differential Pre-

dation Drives the Geographical Divergence in Multiple Traits in Aposematic Frogs." *Behavioral Ecology* 28 (4): 1122–30.

Peng, Xin, C. Zhou, X. Hou, Y. Liu, Z. Wang, Xiaolin Peng, Z. Zhang, R. Wang, and D. Kong. 2018. "Molecular Characterization and Bioactivity Evaluation of Two Novel Bombinin Peptides from the Skin Secretion of Oriental Fire-Bellied Toad, *Bombina orientalis*." *Amino Acids* 50:241–53.

5PM: DARWIN'S FROG

Crump, M. L. 2002. "Natural History of Darwin's Frog, *Rhinoderma darwinii*." *Herpetological Natural History* 9 (1): 21–30.

6PM: RETICULATED HUMMING FROG

Bascoulès, S., and P. Smith. 2021. "Mutualism between Frogs (*Chiasmocleis albopunctata*, Microhylidae) and Spiders (*Eupalaestrus campestratus*, Theraphosidae): A New Example from Paraguay." *Alytes* 38 (1–4): 58–63.

Cocroft, R. B., and K. Hambler. 1989. "Observations on a Commensal Relationship of the Microhylid Frog *Chiasmocleis ventrimaculata* and the Burrowing Thera-

phosid Spider *Xenesthis immanis* in Southeastern Peru."
Biotropica 21 (1): 2–8.

von May, R., E. Biggi, H. Cárdenas, M. I. Diaz, C. Alarcón, V.
Herrera, R. Santa-Cruz, F. Tomasinelli, E. P. Westeen,
C. M. Sánchez-Paredes, J. G. Larson, P. O. Title, M. R.
Grundler, M. C. Grundler, A. R. Davis Rabosky, and D. L.
Rabosky. 2019. "Ecological Interactions between Arthro-
pods and Small Vertebrates in a Lowland Amazon
Rainforest." *Amphibian and Reptile Conservation* 13 (1):
65–77.

7PM: ROSENBERG'S GLADIATOR FROG

Höbel, G. 2008. "Plasticity and Geographic Variation in the
Reproductive Ecology of Gladiator Frogs, Particularly
Hypsiboas rosenbergi." *Stapfia* 88:329–34.

Kluge, A. G. 1981. *The Life History, Social Organization, and
Parental Behavior of* Hyla rosenbergi *Boulenger, a Nest-
Building Gladiator Frog.* Miscellaneous Publications,
Museum of Zoology. Ann Arbor: University of Michi-
gan.

8PM: MATANG NARROW-MOUTHED FROG

Das, I., and A. Haas. 2010. "New Species of *Microhyla* from

Sarawak: Old World's Smallest Frogs Crawl Out of Miniature Pitcher Plants on Borneo (Amphibia: Anura: Microhylidae)." *Zootaxa* 2571:37–52.

Gilbert, K. J., T. Goldsborough, W. N. Lam, F. Leong, and N. E. Pierce. 2022. "A Semi-detritivorous Pitcher Plant, *Nepenthes ampullaria* Diverges in Its Regulation of Pitcher Fluid Properties." *Journal of Plant Interactions* 17 (1): 956–66.

Moran, J. A., C. M. Clarke, and B. J. Hawkins. 2003. "From Carnivore to Detritivore? Isotopic Evidence for Leaf Litter Utilization by the Tropical Pitcher Plant *Nepenthes ampullaria*." *International Journal of Plant Sciences* 164 (4): 635–39.

9PM: TÚNGARA FROG

de Silva, P., C. Jaramillo, and X. E. Bernal. 2014. "Feeding Site Selection by Frog-Biting Midges (Diptera: Corethrellidae) on Anuran Hosts." *Journal of Insect Behavior* 27:302–16. https://doi.org/10.1007/s10905-013-9428-y.

Gray, H. M., D. M. Green, and R. Ibáñez. 2021. "Diurnal Calling in a Nocturnal Frog: Exceptional Calling Activity of Túngara Frogs (*Engystomops pustulosus*) on the Panamanian Island of Taboga." *Herpetologica* 77 (3): 227–31.

Halfwerk, W., M. Blaas, L. Kramer, N. Hijner, P. A. Trillo,

X. E. Bernal, R. A. Page, S. Goutte, M. J. Ryan, and J.
Ellers. 2019. "Adaptive Changes in Sexual Signalling in
Response to Urbanization." *Nature Ecology and Evolution*
3:374–80. https://doi.org/10.1038/s41559-018-0751-8.

Rand, A. S., and M. J. Ryan. 1981. "The Adaptive Significance
of a Complex Vocal Repertoire in a Neotropical Frog."
Zeitschrift für Tierpsychologie 57:209–14.

Ryan, M. J., M. D. Tuttle, and L. K. Taft. 1981. "The Costs and
Benefits of Frog Chorusing Behavior." *Behavioral Ecology
and Sociobiology* 8:273–78.

Tuttle, M. D., and M. J. Ryan. 1981. "Bat Predation and the
Evolution of Frog Vocalizations in the Neotropics."
Science 214:677–78.

Tuttle, M. D., L. K. Taft, and M. J. Ryan. 1982. "Evasive
Behavior of a Frog in Response to Bat Predation." *Animal Behaviour* 30:393–97.

10PM: AFRICAN CLAWED FROG

Black, S., K. Larkin, N. Jacqmotte, R. Wassersug, S. Pronych,
and K. Souza. 1996. "Regulative Development of *Xenopus
laevis* in Microgravity." *Advances in Space Research* 17
(6–7): 209–17.

Tinsley, R. C., and H. R. Kobel, eds. 1996. *The Biology of Xenopus*. London: Clarendon Press.

Cooke, A. S., and T. H. Sparks. 2004. "Population Declines of Common Toads (*Bufo bufo*): The Contribution of Road Traffic and Monitoring Value of Casualty Counts." *Herpetological Bulletin* 88:13–26.

Index

monitor lizards, 42, 93
Morocco, 51
mud-nesting frog, 14, 15–20, 145
mud nests, 16–20, 145
Mungomery, Reg, 41–42

narrow-lid pitcher plant, 132–35
narrow-mouthed frogs, 119. *See also* Matang narrow-mouthed frog; reticulated humming frog; tomato frog; Wandolleck's land frog
Nepenthes ampullaria. See narrow-lid pitcher plant
nest construction: mud-nesting frog and, 16–17; Rosenberg's gladiator frog and, 123, 125–27, 129
New Guinea. *See* Wandolleck's land frog
Noanamá Chocó Indians (Colombia), 56
Noble, Gladwyn Kingsley, 50
North America, 119. *See also* crawfish frog; wood frog
Northern Territory (Australia), 42, 85, 88
number of frog species, x–xi
nuptial pads, 49–50

odontoids, 63, 80, 82
Oriental fire-bellied toad, 103–9, 104
overexploitation, xv, 69, 159

Pacific horned frog, 59–65, 60, 101
"pacman frogs," 19, 64–65
Panama, xiv, 12, 123, 137–41
Papua New Guinea, 134
Paraguay, 16, 120
parental care, xii, 30; African bullfrog and, 79, 81–82; Chile Darwin's frog and, 115; common midwife toad and, 45, 47–48, 50–51; Darwin's frog and, 113–14; horned marsupial frog and, 10–11; Jamaican cave-breeding frog and, 30–31; mud-nesting frog and, 17–18; Rosenberg's gladiator frog and, 127–28; Wandolleck's land frog and, 27, 29–30
peptides, antimicrobial, 107–9
Peru, xi, 61, 117, 119, 120
pet trade, frogs in, 65, 69–70, 146
Phyllobates terribilis. See golden poison frog
pixies, 80, 83
pollution, xv, 159
Puerto Rico, 40

180

EARTH DAY